VIVA VOCE
in
Electrical Engineering

By the same authors

- **Admiralty** : Solved Examples in Electrical Calculations, 3rd ed.
- **Admiralty** : Solved Examples in Electrical Calculations (**Hindi**)
- Viva Voce in Electrical Engineering, 5th ed.
  (With Multiple Choice Objective Type Questions)
- Viva Voce in Electrical Engineering, 4th ed. (**Hindi**)
  (With Multiple Choice Objective Type Questions)
- Basic Electrical Engineering & Electronics, 3rd ed.
- Basic Electrical Engineering & Electronics, 2nd ed. (**Hindi**)

VIVA VOCE
in
Electrical Engineering

D.K. Sharma
P. Sharma

CBS

CBS Publishers & Distributors Pvt. Ltd.

New Delhi • Bengaluru • Chennai • Kochi • Kolkata • Mumbai
Hyderabad • Nagpur • Patna • Pune • Vijayawada

ISBN: 978-81-239-0790-1

Fifth Edition: 2002
Reprint: 2003, 2004, 2007, 2008, 2009, 2010,
2011, 2012, 2013, 2014, 2015, 2016

Published by:
Satish Kumar Jain for CBS Publishers & Distributors Pvt. Ltd.,
4819/XI Prahlad Street, 24 Ansari Road, Daryaganj, New Delhi - 110002
delhi@cbspd.com, cbspubs@airtelmail.in • www.cbspd.com
Ph.: 23289259, 23266861, 23266867 • Fax: 011-23243014

Corporate Office: 204 FIE, Industrial Area, Patparganj, Delhi - 110 092
Ph: 49344934 • Fax: 011-49344935
E-mail: publishing@cbspd.com • publicity@cbspd.com

Branches:
- ***Bengaluru:*** 2975, 17th Cross, K.R. Road, Bansankari 2nd Stage, Bengaluru - 70 • Ph: +91-80-26771678/79 • Fax: +91-80-26771680 E-mail: cbsbng@gmail.com, bangalore@cbspd.com
- ***Chennai:*** No. 7, Subbaraya Street, Shenoy Nagar, Chennai - 600030 Ph: +91-44-26681266, 26680620 • Fax: +91-44-42032115 E-mail: chennai@cbspd.com
- ***Kochi:*** Ashana House, 39/1904, A.M. Thomas Road, Valanjambalam, Ernakulum, Kochi • Ph: +91-484-4059061-65 Fax: +91-484-4059065 • E-mail: cochin@cbspd.com
- ***Kolkata:*** 6-B, Ground Floor, Rameshwar Shaw Road, Kolkata - 700014 Ph: +91-33-22891126/7/8 • E-mail: kolkata@cbspd.com
- ***Mumbai:*** 83-C, Dr. E. Moses Road, Worli, Mumbai - 400018 Ph: +91-9833017933, 022-24902340/41 • E-mail: mumbai@cbspd.com

Representatives:
- Hyderabad: 0-9885175004
- Patna: 0-9334159340
- Vijayawada: 0-9000660880
- Nagpur: 0-9021734563
- Pune: 0-9623451994

Printed at:
J.S. Offset Printers, Delhi

Preface to the Fifth Edition

We are feeling a great pleasure to introduce the 5th edition of this book. We are encouraged by the great response shown by the readers.

This edition of the book covers a vast range of objective and viva voce questions from different angles to make the subject easy and understandable and activity-oriented.

It aims at an integrated development in all ways of listening, reading and writing to make learning most effective. Numerous carefully selected questions are added keeping in view the principle of learning and step-by-step grasping.

It is earnestly hoped that the method of presentation employed in this book will fully serve the needs of the young engineers in electrical engineering.

We will be obliged if the teachers and students point out any shortcomings in this edition for further development.

D.K. Sharma
P. Sharma

Preface to the First Edition

The preparation of this book was undertaken to meet the increasing demand that has been growing for some years for a suitable book of 'Viva voce' in Electrical Engineering;' according to the syllabus of the various Technical Boards. The main aim of this book is to furnish all relevent questions on the subject arranged in a manner to suit all the trainees of technical institutions specially the I.T.I., B.T.C. and Diploma level students of electrical Engineering and Electronics branches. Moreover every chapter in this book has been dealt in a very simple systematic and lucid manner starting from very fundamental concepts. Care has been taken to make the language of this book too simple to understand.

While presenting this book in the hands of readers, it is fervently hoped that the author's new venture will meet the ever increasing demand of not only the students in I.T.I. and Polytechniee but it shall prove a useful hand book to the engineers and candidates preparing for different level interviews in the engineering profession.

I am thankful to M/s C.B.S. Publishers for publishing this book in a very short period, giving a handsome look to present.

The success and popularity of this venture depends upon the liking of the readers whose valuable suggestions are invariably welcome to improve upon the book. Errors might have crept in despite atmost care and author shall be thankful if these are pointed out along with the other suggession if any.

Karkardooma
New Delhi-92

D.K. Sharma
P. Sharma

Contents

CHAPTER 1

Safety Precautions

INTRODUCTION

Q. **How a person gets shock?**

Ans. Whenever current flows through the body of a person, he gets shock.

Q. **What is the general voltage in the house?**

Ans 230 volts.

Q. **What is the lowest recorded voltage, when a person starts feeling some sensation of electric shock?**

Ans. About 30 volts.

Q. **How the safety is ensured?**

Ans. By observing the safety precautions in operations and handling of an electrical appliance.

Q. **How the safety of electrical appliances and electrical wiring is ensured?**

Ans. It can be ensured by three ways—proper insulation, good earthing and by providing a fuse in the circuit.

Q. **What are the precautions you will observe while handling an electrical appliance?**

Ans.
1. Never temper with any electrical appliance or any electrical gear.
2. Make sure that the appliance is properly earthed.
3. The cable insulation should be sound.
4. Never disconnect the appliance by pulling out flexible cable.
5. Before handling a portable appliance make sure that it is disconnected; merely switching off is not enough.

Q. **What are the main precautions to be observed while working on overhead line?**

Ans.
1. Never touch an overhead line conductor unless you are sure that it is dead and properly earthed.
2. Never energize the line of a conductor unless you are sure that all is clear and there is no one working on the line.
3. Never forget to put on safety belt when you are working on a pole or tower or overhead line.
4. Short-circuit the OH line before starting work on the line. Generally a chain is used for this propose if the line is dead.
5. Use rubber gloves.

Q. **What precautions will you observe while charging a battery?**

Ans.
1. Keep proper ventilation.
2. When the cells are being charged, never bring a naked flame near it.
3. While mixing sulphuric acid to water, always add acid to the water that too drop by drop from the sides of the container and not vice versa.

Q. **If the site of working is at a far distance from the place of switch, what precaution will you take?**

Ans. Off the switch and put a plate "Men at work" on the switch so that no one should operate it and also take off the fuses.

Q. **Is it necessary to put off the switch before replacing the blown fuse?**

Ans. Yes, otherwise if the fuse melted because of earth fault or short-circuit, it will again blow off as soon as it is replaced and may cause damage to the operator; so for the sake of safety it should be replaced only after putting the switch off.

Q. **Is there any safety provided by the manufacturer so that the fuse may not be taken out unless the switch is off?**

Ans. Yes, some mechanical arrangement is provided with the cover and base. It will only permit to take out the fuses if the switch is off.

Q. **What precaution should be taken while the ladder is in use?**

Ans. Ladder must be held by another person on the ground so that it may not slip down.

Q. **Is it advisable to tie wire or rope with the electric pole for drying out the clothes etc.?**

Ans. Never. In case of earth fault or leakage, it may give shock.

Q. **Should we throw water in case of the electric fire?**

Ans. Never.

Q. **What precautions should be taken in case of electric fire?**

Ans.
1. First of all put the main switch off if it is on and is within your approach, disconnect the supply immediately.

2. Throw or pour dry sand on the place where the fire took place.
3. Use fire extinguisher and that too after being sure that it is suitable for that purpose.
4. Inform doctor in case of any casualty and to the nearest substation also.

TREATMENT OF ELECTRIC SHOCK

Q. **While a person is in contact with the electric shock, should he be removed by pulling his arm?**

Ans. Never do it.

Q. **How he should be removed from the electric contact?**

Ans. Either by pushing him with the help of a wooden rod or any other dry wood. Or insulate yourself on the dry paper or dry wooden article or any other suitable insulating medium like dry clothes and pull or push him and remove from the contact. Put off the switch immediately.

Q. **In case of electric shock what precautions would you observe?**

Ans.
1. Removal from the contact.
2. Sending for the doctor.
3. Informing about the accident to the nearest substation.
4. Also trying for artificial respiration until doctor comes, if it is needed.

Q. **Should something be given to a person to drink when he is unconscious?**

Ans. Never give anything to drink when the patient is unconscious.

Q. **Should the artificial respiration be stopped as soon as the patient starts rebreathing?**

Ans. No, but it should be continued for some more time.

Q. **In case of OH line, which points should be earthed?**

Ans. According to the Rule No. 90 of IER, all metal supports of OH and metallic fittings attached thereto shall be permanently and efficiently earthed. For this purpose a continuous earth wire shall be provided and securely fastened to each pole and connected with earth ordinarily at four points in every 1.61 km (1 mile).

Q. **What preventive precautions should be taken to avoid electric shock?**

Ans. Proper insulation, proper handling, proper earthing and proper use.

Q. **What are the different methods for artificial respiration?**

Ans. Generally these are :

1. Schafor's method

2. Neilson's method
3. Mouth to nose method.

Q. **How the pressure is applied on the patient in Schafor's method?**

Ans. Keeping both hands on the back of the patient spread your fingers and thumb touching each other. Now lean over the patient forward and apply steady pressure gently.

Q. **How much should be the frequency of operation?**

Ans. It should be 15 times in a minute.

Q. **How long it should be continued?**

Ans. Till the doctor comes or he breaths.

Q. **Where the Neilson's method is adopted?**

Ans. If the burns have taken place on the front position then this method is adopted.

Q. **What is the position of head in this method and how it is done?**

Ans. In this method, the head should fall back, open his mouth and take out the tongue. Place a pillow or his coat rolled up beneath his shoulders so that his head falls back.

Q. **How the arms be kept in the Neilson's method?**

Ans. The arms should be held just below the elbow and draw his arms over his head until horizontal.

Q. **How much time the hands be retained in the horizontal position?**

Ans. For two seconds.

Q. **What should be the frequency of operation?**

Ans. Same, 15 times in one minute.

Q. **What should be the position of the patient in case of mouth to nose method?**

Ans. Lay him down as in case of Neilson's method and his chest upward.

Q. **How this mouth to nose method is done?**

Ans. Take a deep breath of fresh air and blow into the nose of the patient.

Q. **What precautions in this test should be observed?**

Ans.
1. The nose should be cleaned properly.
2. The patient should not be a patient of T.B. or any other infectious disease.

Q. **What should be the frequency of the operation?**

Ans. At least 12 times in a minute.

Q. **Can you give some important causes of accidents?**

Ans. Generally the accidents do not just happen, they are caused. Some of the reasons may be disregard of safety (that includes unawareness, negligence or overconfidence), untidy conditions of workshop, inadequate illumination, improper use of tools etc.

Q. **What type of injury is caused due to rough surface, jagged edge or splinters of sharp or pointed projections?**

Ans. Cuts or abrasions.

Q. **What are ten general safety precautions?**

Ans. General safety precautions can be said as under :

1. Switch should always be put on the positive or live conductor.
2. The interlockings in no case be interrupted unless it is essential and that too if you are authorised to do so.
3. Never tamper the interlockings, e.g., main switch and other safety equipments, because these are for your safety.
4. All the metallic parts of the equipments, installation and machine except current-carrying conductor must be properly earthed.
5. Always try to use insulated tools.
6. Never disconnect a flexible lead just by pulling it.
7. Three-pin plug tops be preferred because it has the earth terminal which is an essential part of electricity.
8. Never energise a line or live wire unless you are sure that everything is clear and there is nobody working on it.
9. Never forget to put on your safety belt when you are working on an overhead line.
10. When you are repairing any appliance or equipment or machine approach should be systematic. Do not proceed haphazardly.

Q. **Should we rush to doctor immediately just after the patient gets shock or start first aid?**

Ans. First attend the patient, start first aid and arrange somebody to rush for doctor.

Q. **Why the use of water is condemned for electric fire?**

Ans. Because water is a conductor of electricity so the risk of electric shock be avoided at every cost.

Q. **How will you replace a blown fuse?**

Ans. First off the main switch, check the condition of melted fuse element. Test the circuit for earth fault, place the wired grip into its housing, cover it and then on the main switch that too standing by a handle side not in front.

Q. **Which types of tools will you prefer to use?**

Ans. Insulated tools.

OBJECTIVE TYPE MULTIPLE CHOICE QUESTIONS

1. The supply voltage used for domestic purposes is ......
   (a) 110-120 V
   (b) 120-130 V
   (c) 220-230 V
   (d) 400-440 V.
2. In industrial work ...... supply is used in India.
   (a) 400 V, 50 c/s
   (b) 220 V, 40 c/s
   (c) 110 V, 50 c/s
   (d) 440 V, 40 c/s.
3. The overhead line should be short-circuited by ...... if it is dead.
   (a) a chain
   (b) copper wire
   (c) the insulated wire
   (d) the safety belt.
4. Use ...... in case of electric fire.
   (a) dry sand
   (b) wet sand
   (c) carbon powder
   (d) water.
5. Add ...... into sulphuric acid to prepare the electrolyte for battery charging.
   (a) ordinary water
   (b) salty water
   (c) distilled water
   (d) dirty water.
6. A person should be removed from the electric contact by ......
   (a) pulling him from his arm
   (b) pulling him from his hair
   (c) pulling him from his wet clothes
   (d) pulling him from his dry clothing and insulating yourself.
7. While preparing an electrolyte ......
   (a) maintain good ventilation
   (b) never bring a naked light near the place
   (c) add acid to water and not vice versa
   (d) maintain good ventilation; never bring a naked light near the place of preparation and add acid to water.
8. While a person is using a ladder it should be ......
   (a) stopped against the wall only
   (b) should be held in position by at least a person standing near the feet of ladder
   (c) should be tightened by a rope
   (d) left free.

9. While disconnecting a portable equipment from the line ……
   (a) never disconnect just by pulling the flexible cord
   (b) disconnect by pulling the cord
   (c) merely switching off is sufficient
   (d) switch off and never disconnect just by pulling the flexible cord.
10. Use …… and …… while working on OH line.
   (a) insulated tools, gloves
   (b) safety belt, gloves
   (c) safety belt, insulated tools and rubber gloves
   (d) safety belt, insulated tools.

CHAPTER 2

Electrician Tools

INTRODUCTION

Q. **Name the tools which are commonly used by an electrician?**

Ans. Insulating pliers, screw driver, cutter, pocker, drill machine, cold chisel, hammer, firmer chisel, line tester, hand saw, etc.

Q. **Is it necessary to use an insulated pliers by an electrician?**

Ans. Yes, because an electrician has to work for electricity, sometimes on the running line also so only the insulated tools should be preferred.

Q. **What cares and maintenances are required for an insulated combination pliers?**

Ans. Don't cut steel wires or any hard substance with the pliers, don't hold much cold or hot substances, don't use it as hammer.

The insulation should be maintained properly intact and should be sound and strong. Oil it when required.

Q. **What are the different sizes of pliers?**

Ans. These are available in length as 15 cm, 20 cm, 25 cm, etc.

Q. **What is a screw driver?**

Ans. It is a tool which is used to tighten or loosen the screws or bolts. These are also available in length like, 6 cm, 15 cm, 25 cm, etc.

Q. **What for a pocker is used?**

Ans. To make the pilot holes for screw tightening.

Q. **Name the tool which is used for cutting the wooden article?**

Ans. Hand saw or tenon saw.

Q. **Can a hand saw be used for cutting conduit pipe?**

Ans. No, hacksaw is used for this purpose.

Q. **Name the tool used for sharpening the hand saw?**

Ans. Triangular file.

Q. **What are the types of the chisels?**

Ans. These are cold chisel which is used for making holes or digging channels in the walls etc., and firmer chisel which is used for wooden articles.

Q. **How the firmer chisels are available in the market?**

Ans. According to the width of the blade.

Q. **What are the different sizes available?**

Ans. It varies from 1.25 cm to 2.54 cm.

Q. **What is mallet and where it is used?**

Ans. It is a hammer made from wooden block and it is used for straightening the metallic sheets, striking the firmer chisel and fitting the electric machines, like motor etc.

Q. **What for the rawl plug tool is used?**

Ans. It is made of steel and is used for making holes in the stones, walls or on concrete wall, etc.

Q. **What for stock and die set is used?**

Ans. It is used the making threads on the conduit pipes.

Q. **Can you use the jaws of different sizes?**

Ans. Yes.

Q. **How are the sizes taken and what are common?**

Ans. The sizes are taken according to the diameter of the jaws. These are available in ½″, ¾″, 1″ and 1½″ sizes etc.

Q. **What precautions should be taken while threading a conduit pipe?**

Ans. Apply oil for lubrication, move in both directions forward and backward so that the threading is easy and fine. Tight slowly, so that heavy cut may not disturb the joint of pipe, after that clean the threading.

Q. **Name the file used on wooden article?**

Ans. Rasp cut file.

Q. **What are the general types of hammer?**

Ans. Ballpein, straightpein, crosspein.

Q. **Name the vice used for drawing wires of the line?**

Ans. Draw vice.

Q. **Name the tool used for measuring the gauge of conductors?**

Ans. S.W.G.

OBJECTIVE TYPE MULTIPLE CHOICE QUESTIONS

1. The gauge of a bare conductor is measured by ......
   (a) steel ft. rule (b) steel tape
   (c) S.W.G. (Standard Wire Gauge)
   (d) Four fold wooden ft. rule.
2. Which tool is used for pulling, cutting, twisting and wrapping purposes?
   (a) Screw driver (b) Insulated combination pliers
   (c) Side cutter (d) Gas pliers.
3. Which file is used for sharpening the teeth of a hand saw?
   (a) Rasp cut file (b) Flat file
   (c) Triangular file (d) Round file.
4. Which tool is used for hammering the nails?
   (a) Screw driver (b) Mallet
   (c) Pliers (d) Hammer.
5. Firmer chisels are available in the market according to the ......
   (a) length of the blade (b) width of the blade
   (c) thickness of the blade (d) complete length of the tool.
6. For threading the conduits ...... is used.
   (a) stock and die set (b) internal die set
   (c) hacksaw (d) spanner set.
7. Which tool should be used for tightening the screws?
   (a) Firmer chisel (b) Screw driver
   (c) Combination pliers (d) Pocker.
8. Which tool should be used for making holes for gitties?
   (a) Screw driver (b) Firmer chisel
   (c) Cold chisel (d) Pocker.
9. Draw vice is used for tightening the ......
   (a) straight joint of P.V.C. wire. (b) O.H. conductors.
   (c) pipe for threading purpose. (d) bolt for machine foundation.
10. The use of combination pliers be discouraged for ......
    (a) cutting the wires (b) twisting the wire
    (c) skinning the conductors (d) hammering the objects.

CHAPTER **3**

Fundamental Units

INTRODUCTION

Q. **How many system of units are generally used in Engineering?**

Ans. Generally four.

Q. **Name the different systems of units?**

Ans. C.G.S., F.P.S., M.K.S. and S.I. system.

Q. **What is C.G.S. system?**

Ans. The system of unit in which length is expressed in centimetre, weight or mass in grams and time in seconds.

Q. **Name the system in which length is measured in foot, weight in pounds and time in seconds?**

Ans. F.P.S. system.

Q. **What is MKS system?**

Ans. It is the system of engineering units in which length is measured in metre, weight in kilogram and time in seconds.

Q. **Define MKSA system?**

Ans. It has four units—the length, mass, time and the current measured in metres, kilograms, seconds and amperes, respectively.

Q. **What is S.I. system of units?**

Ans. It is nothing but the rationalized MKSA system. It has six units—the length, the mass, the time, the current, the thermodynamic temperature and luminous intensity measured in metres, kilograms, seconds, amperes, kelvin and candela, respectively.

Q. **Define matter?**

Ans. Matter is that which occupies some space, has some weight and can be felt by one or more of our senses.

Q. **What is mass?**

Ans. Mass is the quantity of matter contained by a body.

Q. **What is the difference between mass and weight?**

Ans. Mass of a body does not change whereas the weight of body changes from place to place. A body will have different weight at different places.

Q. **What is speed?**

Ans. It is the rate of change of motion of a body in any direction. It is represented in km/hr, cm/sec etc.

Q. **What is difference between speed and velocity?**

Ans. In speed the direction is not mentioned whereas in velocity the direction is given. The speed is a scalar quantity and velocity is a vector quantity.

Q. **What is acceleration?**

Ans. It is the rate of change of velocity. The unit is m/sec^2 or cm/sec^2.

Q. **What is retardation?**

Ans. The negative acceleration is known as retardation.

Q. **What is the unit of retardation?**

Ans. Same as km/hrs^2 or cm/sec^2.

Q. **What is the force?**

Ans. Force can be defined by the effects it exhibits. So the force is that which changes or tends to change the state of rest or motion of a body.

Q. **What is the unit of force?**

Ans. It is Newton in M.K.S. system.

Q. **What is one Newton?**

Ans. It is the force which produces the acceleration of one metre per sec^2 in one kg weight.

Q. **What is work?**

Ans. When a force is applied on a body and the body moves through a certain distance, work is said to be done.

OR

It can be defined as, "work is said to be done when a force acting on a body causes it to move."

Q. **If the force acting and distance moved is given, how the work done is calculated?**

Ans. Work done = Force × Distance.

Q. **What do you understand by the term torque?**

Ans. The torque is defined as the turning and twisting moment of a force about its axis.

Q. **How the torque is measured?**

Ans. It is measured by the product of the force and the perpendicular distance from the line of action to the body.

Q. **What is the unit of torque?**

Ans. It is Newton-metre.

Q. **Define heat and calorie?**

Ans. Heat is the form of energy which produces the sensation of warmness or coldness. Its unit is calorie.

Q. **What is one calorie?**

Ans. One calorie is that much amount of heat which is required to raise the temperature of one gram of water through one degree centigrade.

Q. **What is temperature?**

Ans. The temperature is the degree of hotness or coldness of a body. It is measured by the thermometer.

Q. **How many general types of thermometer are there?**

Ans. These are Centigrade, Fahrenheit, Reaumer and Kelvin.

Q. **Can you change a temperature of one scale to the another scale?**

Ans. Yes.

Q. **How this can be changed or transferred?**

Ans. By using the formula

$$\frac{C}{5} = \frac{F - 32}{9} = \frac{R}{4}$$

Q. **What is the boiling point of water in °C, °F and °R?**

Ans. It is 100°C, 212°F and 80°R respectively.

Q. **What is the temperature which is same on °F and °C?**

Ans. –40°.

Q. **What is the freezing point of water in °F?**

Ans. 32°C.

Q. **What is power?**

Ans. It is the rate of doing work.

Q. **What are the units of power?**

Ans. In mechanical engineering the unit of power is H.P., in electrical engineering it is Watts. The bigger units are kW and MW.

Q. **What is one h.p.?**

Ans. If any agent does the work at a rate of 550 ft lb/sec or 33000 ft lb/minute then one H.P. is said to be consumed.

Q. **How much watts are equal to one h.p.?**

Ans. 1 HP = 746 Watts.

Q. **What are the formulae for calculating power?**

Ans. $W = (\text{Current})^2 \times \text{Resistance}$

$$= \text{Voltage} \times \text{Current}$$

$$= \frac{(\text{Voltage})^2}{\text{Resistance}}$$

Q. **If a heater has 400 Ω resistance and drawing a current of 2 Amp., what is the power consumed?**

Ans. 1600 Watts.

Q. **If two resistances one of more value and other of low value are connected in parallel across same voltage, which one will consume more power?**

Ans. Low value resistance.

Q. **What is energy?**

Ans. It is capacity of doing work.

Q. **What is the unit of electrical energy?**

Ans. It is watt sec or watt hour, kilowatt hour.

Q. **What is one kilowatt hour?**

Ans. $1\ \text{kWh} = \dfrac{\text{Volt} \times \text{Current} \times \text{Time (hrs)}}{1000}$

Q. **How energy and power are related?**

Ans. Energy = Power × Time.

Q. **If a bulb of 100 W lights for 20 hours how much units of electricity is consumed?**

Ans. 2 units.

Q. **What is the law of conservation of energy?**

Ans. It states that the total amount of energy of this universe is constant, thought it may change its state from one form to another.

Q. **What are the Newton's laws of motion?**

Ans. These are :

1. Every body continues in its state of rest or uniform motion in a straight line unless it is compelled by some external force to change its state.
2. The rate of change of momentum of a body is proportional to the impressed force and takes place the same direction in which it is applied.
3. To every action there is equal and opposite reaction.

Q. **A man gets jerk when the vehicle starts. Which law is applicable?**

Ans. First law of motion.

Q. **To accelerate the speed of a vehicle which law is enforced?**

Ans. Second law.

Q. **Why a ball after striking the wall comes back to the agent.**

Ans. According to Newton's third law of motion, to every action there is equal and opposite reaction.

OBJECTIVE TYPE MULTIPLE CHOICE QUESTIONS

1. Which are the fundamental units?
   (a) C.G.S., F.P.S. and S.I. (b) M.K.S., M.K.S.A. and C.G.S.
   (c) F.P.S., M.K.S.A. and M.K.S. (d) C.G.S., F.P.S. and M.K.S.
2. In C.G.S. system of units the length and mass are expressed in ......
   (a) metre and kilogram (b) ft. and kg
   (c) km and gm (d) cm and gm.
3. In M.K.S. system of units the length, mass and time are expressed in ......
   (a) ft, kg and sec. (b) metre, kg and sec.
   (c) cm, gm and sec. (d) ft., lb and sec.
4. Apart from these three mass, length, and time, other three physical quantities in the international system (S.I.) of units are ......
   (a) current, voltage and resistance
   (b) current, voltage and luminous intensity
   (c) current temperature and luminous intensity
   (d) current, temperature and voltage.
5. The S.I. derived units in mechanical engineering are ......
   (a) newton, joules, pascal, watts and hertz
   (b) newton, joules, henry, watts, pascal and hertz
   (c) newton, joule, coulombs, pascal, watt and h.p.
   (d) None of the above
6. If the voltage fed to an electric circuit is reduced to half, the power consumed will reduced to
   (a) 25% (b) 100%
   (c) 50% (d) 75%
7. An electric kettle takes 20 minutes to boil some quantity of water, to boil the same water in 5 minutes ......
   (a) increase the length of resistance of heating element.
   (b) increase the diameter of the wire of heating element.
   (c) increase the insulation resistance.
   (d) increase the capacity of the plug shoe.
8. One horse power is equal to ......

(a) 746 watts
(b) 750 watts
(c) 740 watts
(d) 756 watts.

9. One B.O.T. unit is equal to ......
(a) 1000 watts
(b) 100 watts
(c) 10 watts
(d) 10000 watts

10. One megaohm is equal to ......
(a) $10^6\ \Omega$
(b) $10^3\ \Omega$
(c) $10^8\ \Omega$
(d) $10\ \Omega$

11. Choose the correct unit.
(a) Resistance is measured in ... (Ohm, Amp., Volts)
(b) Current is represented in ... (kWh, Amp., Ohm)
(c) The unit of power is ... (Amp, Kelvin, Watts)
(d) The quantity of electricity is given in ... (Coul, Amp., Volts)
(e) ... is the unit of electric energy (kWh., KVARH, H.P.)

12. Name the instrument used for measuring
(a) Current ... (Ammeter/Voltmeter)
(b) Voltage ... (Ammeter/Voltmeter)
(c) Frequency ... (Frequency meter/Ammeter)
(d) Power factor ... (P.f. meter/Frequency meter)
(e) Power ... (Power factor meter/Wattmeter).

13. State true of false.
(a) E.m.f. represents electromotive force.
(b) P.D. represents potential difference.
(c) The property that helps the current to flow is known as conductance.
(d) Current is measured in Coulombs/sec.
(e) The unit of temperature is calorie.

14. Choose the correct instrument.
(a) Energy is measured by ... (Energy meter, Wattmeter, Ammeter)
(b) Temperature is measured by ... (Wattmeter, Thermometer, Ammeter)
(c) Insulation resistance is measured by ... (Ohm meter, Ammeter, Megger)
(d) Earth resistance is measured by ... (Earth tester, Megger, Line tester)
(e) Voltage is measured by ... (Voltmeter, Ammeter, Ohmmeter)

15. In C.G.S. system of units, the Stoke is the unit of ......
(a) dynamic velocity
(b) kinetic energy
(c) kinematic viscosity
(d) None of the above.

CHAPTER 4

Electrical Accessories

INTRODUCTION

Q. **What do you understand by the electrical accessories?**

Ans. The materials used in wiring or in electrical engineering like switches, plugs, holders, ceiling rose, fuses, etc., are known as electrical accessories.

Q. **Define the controlling accessories.**

Ans. The accessories which are used to control the flow of current are called controlling accessories, namely the switches, isolators and main switches etc.

Q. **Name the different switches?**

Ans. Single-way switch, two-way switch, intermediate switch, double pole double-throw switch, double-pole iron clad main switch, double pole iron clad main switch with fuses, triple pole main switch with and without fuses, knife switch and bed switch, etc.

Q. **Why the single way switch is named so?**

Ans. Because it controls the flow of current in one way or one line i.e. one circuit only.

Q. **Why the two way switch is called the two way switch?**

Ans. Because it controls the flow of current in two ways or two circuits.

Q. **How many terminals are there in a two way switch?**

Ans. It has four terminals out of which two are short-circuited with a thick strip or three terminals out of them one terminal with long strip on one side so that the continuity can be maintained for two circuits (ways).

Q. **Where the two-ways switches are generally used?**

Ans. In staircase points.

Q. **Where intermediate switch is used?**

Ans. When a point is to be controlled from more than two places.

Q. **Which switch is used to control a bell point?**

Ans. Bell push.

Q. **How these switches are connected?**

Ans. In series with the appliance.

Q. **Why the double pole iron clad main switch is named so?**

Ans. Because it controls both phase and neutral or +ve and –ve wires simultaneously.

Q. **Why generally in house wiring the main switch with neutral link is preferred?**

Ans. The fuse is always connected on the phase wire and if by any reason fuse melts, then the phase is disconnected not the neutral thus reducing the danger of shock.

Q. **What are the ratings of single-way switch?**

Ans. 5 A, 250 V and 15 A, 250 V.

Q. **Can you use a 5 A, 250 V switch for 15 A, 250 V?**

Ans. No.

Q. **Why 5 A switch cannot be used for 15 A, 250 V?**

Ans. Because their terminals and strips are designed only for 5 A and if current will be more than 5 A the strip will melt out and the switch will be damaged.

Q. **Can you recognise a tumbler switch (when it is covered) whether it is a single way switch?**

Ans. Yes seeing from back, if it has two terminals then it is single way and if more then two way.

Q. **What is the name of the switch which is generally used in hanging position near the bed?**

Ans. Bed switch.

Q. **What are the different types of triple pole main switches?**

Ans. Triple pole iron clad main switch (TPIC) with and without fuses and triple pole double throw iron clad main switch.

Q. **Where a T.P.I.C. main switch is used?**

Ans. On three phase line.

Q. **Where D.P.I.C. is used?**

Ans. In single phase or D.C. supply.

Q. **What are the different types of holder?**

Ans. These are batten holder, pendant holder, bracket holder, angle holder, screw type holder, tube holder, etc.

Q. **Name the holder which is used to hang the light.**

Ans. Pendant holder.

Q. **What do you understand by the batten holder?**

Ans. The holder which is used on the round block or on the flat surface is called the batten holder.

Q. **Upto which wattage the batten holders are used?**

Ans. 200 W.

Q. **Above 200 W lamps generally which type of holders are used?**

Ans. Screw types.

Q. **What is rating of the batten holder?**

Ans. 5 A, 250 V.

Q. **What is the angular relationship between the groves of the holder and the pins?**

Ans. Both are at 90° to each other.

Q. **Where the ceiling rose is used?**

Ans. This is used to take out the supply for the pendant light, fan, fluorescent tube, etc.

Q. **Name the material of which the ceiling rose is made.**

Ans. Bakelite or plastic.

Q. **What are the types of the ceiling rose?**

Ans. Two plates and three plates.

Q. **Name the accessory which is used to take out supply for table fan or any other portable equipments?**

Ans. Two or three pin plug sockets.

Q. **What is the difference between two pin and three pin plug socket?**

Ans. In two pin plug socket there are only two terminals, one positive or phase and other negative or neutral. In three pin plug there are three terminals one positive or phase, other negative or neutral and the third the earth.

Q. **Name the accessory which is used to take supply for a number of appliances from a single point?**

Ans. Multiplug or adopter.

Q. **What are the safety accessories?**

Ans. The accessories which check the flow of heavy current in the circuit are known as the safety accessories.

Q. **What is a distribution box?**

Ans. It is a unit in which a number of kitket fuses are employed for different circuits. These are used when the load is more. Generally it is covered in a metallic box.

Q. **What is the plug top?**

Ans. The accessory by which supply is taken for the appliance from the plug socket.

Q. **How many pins are there in 2-pin and 3-pin plug top?**
Ans. 2 and 3 respectively.

Q. **Where the earth terminal is provided?**
Ans. Earth point is provided on the upper side.

Q. **The earth terminal is made bigger and long in size, why?**
Ans. Because before connecting the appliance to phase and neutral the earth continuity is maintained first, as the pin is kept long enough. The area is bigger because in case of the short circuit, current will be certainly more than the current rating of the appliance.

Q. **How many wattage or load is recommended in one circuit of light?**
Ans. It is 800 watts or ten points.

Q. **What is the load in case of one circuit of power?**
Ans. It is 2000 watts per circuit or two points.

Q. **How many types of connectors are there?**
Ans. One way, two ways and three ways.

Q. **Where the connectors are used?**
Ans. These are used to increase the length of the wire without jointing.

Q. **Of which material these are made?**
Ans. Porcelain and brass sleeve with holes and screws.

OBJECTIVE TYPE
MULTIPLE CHOICE QUESTIONS

1. Switch always be installed on ......
   (a) neutral wire (b) earth wire
   (c) phase wire (d) None of the above.
2. Which are the controlling accessories?
   (a) holder and plug
   (b) single pole switch and bed switch
   (c) ceiling rose and cutout
   (d) connectors and brackets.
3. Intermediate switch is used to control a lamp from ......
   (a) one place (b) more than two places
   (c) individual controlling (d) None of the above.
4. Ceiling rose is used to take supply from
   (a) portable equipment (b) fluorescent tube
   (c) a heater of 2000 W (d) an electric iron.
5. The earth terminal of a 3-pin socket is made of the ...... than other two.

(a) same size (b) smaller in size
(c) long and thin (d) larger in dia and length.

6. Which are the holding accessories?
   (a) Switch, plug, socket and holders.
   (b) Holder, kitket fuse, connector and cutouts.
   (c) Batten holder, lamp holder, pendant holder and bracket holders.
   (d) Fuse, indicator, connector and switch.
7. The followings are the protective accessories :
   (a) switch and plug (b) connector and fuse
   (c) cutout and fuse (d) switch and fuse
8. How many maximum number of light points can be connected in one electric domestic circuit?
   (a) 8 (b) 10 (c) 12 (d) 15
9. How many maximum number of points in case of mixed load (light and power) can be installed in one electric circuit?
   (a) 10 (b) 8 (c) 12 (d) 15
10. How many power sockets can be put in one domestic power circuit?
    (a) 2 (b) 1 (c) 4 (d) 6
11. Which DPIC main switch is preferred in domestic A.C. light circuit?
    (a) With fuses. (b) With neutral link.
    (c) Without fuse. (d) None of the above.
12. In TPIC main switch neutral link is provided
    (a) to take supply for lightning and power balanced or unbalanced load.
    (b) To take supply for lamp loads only.
    (c) To take supply for power plugs only.
    (d) To take supply for three phase and single phase power and lightning loads.
13. Fill in the blanks :
    (a) Switch should always be provided on ......
    (b) Earth wire continuity should be maintained throughout the ......
    (c) Pendant holder is used for the lights in ...... positions.
    (d) To distribute the electric circuit for more than one circuit ...... should be preferred.
    (e) Neutral wire should be taken to every individual circuit via distribution box by connecting to the ......

CHAPTER 5

Atomic Theory

INTRODUCTION

Q. **What is an element?**

Ans. Elements are those which are made by nature and all the atoms of an element have got the same properties as that of the same element.

Q. **What is a compound?**

Ans. A compound is a mixture of two or more than two elements in a fixed proportion viz., H_2O, SO_2, NaCl etc.

Q. **What is a mixture?**

Ans. A mixture is a collection of two or more than two elements in any proportion.

Q. **What do you understand by molecule?**

Ans. The molecule is the smallest possible portion of a compound having the same properties as that of the compound.

Q. **What is atom?**

Ans. Atom is the further decomposition of the molecule and does not have the same properties as that of the molecule. Secondly the atom is the smallest portion of an element and has the same properties as that of the element.

Q. **What is nucleus?**

Ans. It is the central core of an atom.

Q. **What are the main particles in an atom?**

Ans. These are mainly three—the electrons, the protons and the neutrons.

Q. **What charge does electrons have?**

Ans. The electrons are negatively charged particles.

Q. **What is the charge of proton?**

Ans. Protons are positively charged particles.

Q. **Neutrons have which charge?**

Ans. The neutrons are neutral; these are not having neither positive nor negative charge.

Q. **Which is lighter in weight : electron or proton?**

Ans. Electron is lighter in weight.

Q. **How many times the proton is heavy?**

Ans. Proton is heavier than electron by about 1840 times.

Q. **Which is heavy : the proton of neutron?**

Ans. Neutron is heavier than proton, but these are taken as having about equal masses.

Q. **What is the mass of an electron?**

Ans. It is about 9.11×10^{-31} kg.

Q. **What is mass of a proton?**

Ans. It is about 1.67×10^{-27} kg.

Q. **What is the charge on an electron?**

Ans. The electron is having 1.602×10^{-31} coulomb of charge.

Q. **How many electrons are equal to one coulomb charge?**

Ans. One coulomb charge is approximately the charge of 6.28×10^{18} electrons.

Q. **What is the normal charge of an atom in normal stage?**

Ans. In normal conditions, the protons and electrons are equal and electrically the atom is supposed to be neutral.

Q. **What is a free electron?**

Ans. The free electron is the electron which revolves in the outermost unsaturated orbit of an atom.

Q. **What is the unsaturated orbit?**

Ans. The orbit which is not satisfied or not having that much number of electrons which it should possess.

Q. **What is saturated orbit?**

Ans. The orbit which acquires that much electrons as it should have.

Q. **How the electrons are arranged in a specific orbit?**

Ans. These are having a definite relationship. If N is the number of orbit, then the electrons in a particular orbit can be calculated as $2N^2$.

Q. **Which way the electron, proton, neutron etc. are bound?**

Ans. Similarly as the solar system.

Q. **Sun is supposed to be the centre core of the movement of the solar system, here in an atom which is that source?**

Ans. Nucleus.

Q. **How many velocities are there in a moving electron?**

Ans. It has both the linear and circular velocities.

Q. **What is current?**

Ans. The drifting of electrons in a systematic way is called as electric current.

Q. **Why heat is produced in a current-carrying conductor?**

Ans. Because of the collision of electrons.

Q. **How will you divide the substance electrically?**

Ans. These can be classified electrically as conductor, insulator and semi-conductor.

Q. **What is a conductor?**

Ans. The substances in which with a small amount of electric force, drifting of electrons is obtained, are called conductors.

Q. **What is insulator?**

Ans. The substances in which the drifting of electrons is not possible and if at all, the force is increased tremendously, a little and very little amount of electrons are moved, those substances are called as the insulators.

Q. **What are semiconductors?**

Ans. These are not the good conductors and normally behaves like an insulator, but with the application of a small voltage the drifting of electrons is achieved, those substances are called as semiconductors.

Q. **Where the semiconductors are used?**

Ans. In radio and television, I.C. and in other electronic devices.

Q. **Does a proton take an active part in the flow or transfer of current?**

Ans. No, the proton does not take an active part in the flow or transfer of electric current.

Q. **Can you give a rough idea about how many times the diameter of an electron and proton are?**

Ans. The electron is three times larger in diameter than the proton.

Q. **In normal condition in an atom how the protons and electrons are related?**

Ans. The electrons are equal to protons.

Q. **Whom do you call chemically inert elements?**

Ans. The atoms whose outermost orbit is completely filled, they are inactive, these elements are called inert elements.

Q. **Whom do you call valence electrons?**

Ans. The electrons in the outermost orbit (valence shell) are called valence electrons.

Q. **How many free electrons are there in the outermost orbit of a semiconductor?**

Ans. Generally semiconductors have valence shell containing four electrons.

Q. **Is there any relationship between the number of free electrons and conduction?**

Ans. Yes, the more free electrons a material has the better it will conduct.

Q. **How an insulator is useful for electricity?**

Ans. An insulator restricts or prevents the flow of current through unwanted paths.

Q. **Are all the conductors wires?**

Ans. All the wires are conductor of electricity but all the conductors are not necessarily the wires viz. a flat busbar is a conductor of electricity but is not termed as wire.

Q. **What is considered about a cable? What do you understand by a cable in general?**

Ans. A cable consists of two or more conductors inside a single insulated covering.

Q. **Which types of conductors are used in a cable?**

Ans. The conductors in a cable may be bare or insulated.

Q. **In electrical engineering common measure of wire diameter is in which system?**

Ans. The common measure is in gauge number.

Q. **How will you classify the materials according to electrical characteristics?**

Ans. Electrically the materials can be viewed and classified as conducting materials, insulating materials and semiconductor materials.

Q. **Can you give some examples of conducting materials?**

Ans. Copper, aluminium, silver, gold, mercury etc.

Q. **Oil, mica, micanite, dry air are the example of which category of the materials?**

Ans. The insulating materials.

Q. **Give some examples of semiconductor materials?**

Ans. Semiconductor materials are carbon, antimony, bismuth etc.

Q. **What should be the good characteristics of the conducting materials?**

Ans. It should have low resistance, good mechanical strength, malleable, easily jointable, easily available and cheap.

Q. **Can you give some important applications of copper in the electrical engineering field?**

Ans. Copper is basically and principally used for (a) transmission, distribution lines and traction collecting wires; (b) windings of generators, motors, transformers; (c) castings, resistance wires as copper alloys; and (d) certain specially designed magnetic materials etc.

Q. **Generally which substances and preferred for manufacturing copper alloys for special purposes?**

Ans. Cadmium, chromium, silver or tellurium in certain percentages.

Q. **Which impurities are added for making copper-based resistance materials?**

Ans. Constantan, German silver, manganin, and monel metals are used for manufacturing copper-based resistance materials.

Q. **Which materials are used for heavy-duty contacts?**

Ans. Silver and palladium are used for heavy-duty contacts in electrical engineering.

Q. **Can you tell some specially manufactured alloys which are copper-based for preparing permanent magnetic materials?**

Ans. These are copper-nickel-iron (Co-Ni-Fe) alloy and copper-nickel-cobalt with certain proportionals.

Q. **What is major applications of aluminium conductors in electrical engineering?**

Ans. The major application is the ACSR conductors, i.e. the Aluminium Conductors Steel Reinforced (Standard) conductors.

Q. **Can you give some special properties of aluminium instead of conductivity?**

Ans. Resistance to corrosion, non-magnetic and less weight.

Q. **Which combination is suitable for manufacturing busbar casings?**

Ans. Aluminium alloy with 5 to 12% silicon is used for making busbar casings.

Q. **Why it is used for busbar casings?**

Ans. It is light in weight, non-magnetic, easily available and cheap.

Q. **What are the different types of insulator depending upon their stability?**

Ans. The insulators are of three types according to their state of stability. These are solid or hard insulator, the flexible insulator, liquid insulator, and gaseous insulators.

Q. **What are hard insulators?**

Ans. These are hard in nature and are marble, porcelain, glass etc. These insulators lose their dielectric strength as soon as these are deshaped or cracked.

Q. **Whom do you call flexible insulator?**

Ans. The insulators which are not hard and are flexible called flexible insulators such as rubber, cotton and paper etc.

Q. **Under which category of insulators the oils, varnish etc. are covered?**

Ans. The liquid insulator.

Q. **Which type of insulator is the nitrogen gas?**

Ans. It is a gaseous insulator.

Q. **What do you understand by the electric wire?**

Ans. It is an electrical conductor provided with a suitable insulating covering for safe accessibility and to prevent the leakage of current.

Q. **Are these conductor always insulated?**

Ans. Not necessary, they may be insulated or bare.

Q. **Where the base conductors are used?**

Ans. In overhead wiring (street) and in earth continuity conductors.

Q. **As by seeing the winding wire which is used for compressor winding, is it not the bare conductor?**

Ans. It is not precisely bare. It has a covering of good insulation known as superenamel, but it appears to be bare where it is not.

Q. **What are the types of wire generally used for electrical work?**

Ans. These are VIR wire, CTS wire or TRS wire, PVC wire, flexible wire, LCC wire, and enameled wire.

Q. **Generally which types of wires are used in electrical engineering?**

Ans. These are V.I.R. wire, C.T.S. wires or TRS wire, P.V.C. wire, flexible wire and superenameled copper or aluminium wire.

Q. **What is electricity?**

Ans. It is a kind of energy.

Q. **How will you define it?**

Ans. The flow of electrons is known as electricity or an electric current.

Q. **What are the other different types of energy?**

Ans. These can be understood as mechanical, heat, light, sound, chemical etc.

Q. **How many types of electricity is?**

Ans. It is of two types—static and dynamic.

Q. **What is static electricity?**

Ans. Electricity at rest is known as static electricity. It cannot flow from one place to the another.

Q. **What is dynamic electricity?**

Ans. It is also known as current electricity. Electricity in motion. It can be transmitted from one place to the other.

Q. **Name some applications of electricity?**

Ans. It is used in lighting, heating, magnetic, running of motor, welding, battery charging, electroplating, electrotyping, telephone, telegraphic and radio works, production of special rays etc.

Q. **What are the types of current?**

Ans. The current is of two type : (i) Direct current (D.C.), and (ii) Alternating current (AC).

Q. **What is D.C.?**

Ans. The current which has same magnitude, direction and does not change with the time unless the switch is off.

Q. **What is pulsating D.C.?**

Ans. The unidirectional current which changes only its magnitude not the direction, is known as pulsating D.C.

Q. **What is A.C.?**

Ans. The current which changes its magnitude and direction after a fixed interval of time is known as A.C.

Q. **Which type of current is used in radio, electroplating, battery charging?**

Ans. D.C.

Q. **Now-a-days which type of current is commonly used?**

Ans. A.C.

Q. **Mostly generated electricity is of A.C. or D.C. nature?**

Ans. A.C.

Q. **What is volt?**

Ans. It is a unit of measuring the electric pressure.

Q. **What is voltage?**

Ans. It is the electric pressure measured between two points, for example between live and neutral, phase and phase.

Q. **What is the voltage generally used for domestic purposes in India?**

Ans. 230 volts, 50 c/s.

Q. **How the voltages are classified?**

Ans. The voltages are classified as
Low voltage up to 250 V.
Medium voltage 251-650 V.
High voltage 651 V-33 kV
Extra high voltage above 33 kV.

Q. **Define EMF?**

Ans. It is the force which tends to move the electrons in any closed circuit.
or
The force which compels the electrons to move or complete its path. Its unit is volts.

Q. **Can you measure the EMF?**

Ans. No, it cannot be measured but can be calculated.
EMF = PD + Voltage drop inside the source of supply.

Q. **Why the e.m.f. cannot be measured?**

Ans. Because the instrument with which it is to be measured will consume some energy or say some voltage will be dropped which will not come in appearance.

Q. **The voltage measured on the terminal of a battery is E.M.F. or P.D.?**

Ans. P.D.

Q. **What is current?**

Ans. The flow of electrons.

Q. **Which meter will you use to measure the current?**

Ans. Ampere meter.

Q. **What is one ampere?**

Ans. If a voltage of one volt is applied to a resistance of one ohm then one ampere current is said to flow.

Q. **What is one international ampere?**

Ans. It is that steady current which when passed through a solution of silver nitrate ($AgNO_3$) deposits silver at the rate of 0.001118 gm/sec. That current is known as one international ampere.

Q. **What is one international ohm?**

Ans. It is the resistance offered by a conductor of mercury column, 14.4521 gm in weight having a length of 106.3 cm, of uniform area of cross-section at the melting ice temperature.

OBJECTIVE TYPE MULTIPLE CHOICE QUESTIONS

1. The matter can be defined as ......
   (a) the rate of doing work
   (b) the substance which occupies space have some weight and can be felt by our senses.
   (c) having constant mass and weight
   (d) having the constant gravitational pull irrespective of the mass.
2. The combination of two or more elements in any ratio is called ......
   (a) mixture (b) compound
   (c) liquid (d) None of the above
3. The smallest particles of a compound which have got the same properties as that of the compound is called as ......
   (a) molecule (b) atom
   (c) newton (d) ion.
4. The smallest particle of an element is known as ......

(a) atom (b) molecule
(c) nucleus (d) electron

5. The further sub-division of a molecule is known as ......
(a) atom (b) element
(c) free electron (d) orbit.

6. The atom is composed of ......
(a) electrons only (b) protons only
(c) neutrons only (d) electrons, protons and neutrons

7. The number of electrons in an atom are equal to ......
(a) neutrons (b) protons
(c) the atomic number of the substance
(d) None of these

8. The cation has ......
(a) more number of electron than protons
(b) more number of protons than electrons
(c) equal number of protons and neutrons
(d) equal number of protons, neutrons and electrons.

9. The positively charged body has ......
(a) less number of electrons than protons
(b) less number of protons than electrons
(c) equal number of protons and electrons
(d) less number of electrons than neutrons.

10. The charged atom is called ......
(a) molecule (b) ion
(c) nucleus (d) orbit.

11. The electric current is constituted because of the flow of ......
(a) fixed electrons (b) free electrons
(c) neutrons (d) protons

12. The mass of the electron is ......
(a) 1.602×10^{-31} kg (b) 6.24×10^{-18} kg
(c) 9.11×10^{-31} kg (d) 1.371×10^{-31} kg

13. The electron has a charge equal to ......
(a) 1.602×10^{-31} coulomb (b) 1.63×10^{-31} coulomb
(c) 1.66×10^{-31} coulomb (d) 1.673×12^{-27} coulomb

14. The material differs from one another in the number of ...... in their atoms.
(a) protons (b) neutrons
(c) ions (d) electrons

15. The proton is ...... times more than the mass of one electron.
    (a) 1836 (b) 1736
    (c) 1386 (d) 1639.
16. Which element is a good conductor having
    (a) its outermost orbit in saturated state
    (b) its outermost orbit in unsaturated state
    (c) its outermost orbit unsaturated with one free electron
    (d) None of the above.
17. The atomic weight of an element is based on ......
    (a) free electron (b) protons
    (c) neutrons (d) None of the above.
18. Which is the best conductor ......
    (a) iron (b) copper
    (c) silver (d) aluminium
19. Which is the best insulator?
    (a) Copper (b) Dry air
    (c) Moist air (d) Mica
20. Which of the conductive substance has the highest resistivity?
    (a) Carbon (b) Tungsten
    (c) Nickel (d) Iron
21. The kinetic energy of bonded electron is ...... to the kinetic energy of a free electron.
    (a) equal (b) greater than
    (c) less than (d) not applicable.
22. Kinetic energy of a free electron in a uniform field ......
    (a) increases as square of the mass of electron
    (b) increases as the square of the wave number
    (c) is constant for all wave number
    (d) None of the above.
23. The resistance of a material decrease if ......
    (a) the number of neutrons increased
    (b) the number of free electrons available become more
    (c) the temperature of the material is increased
    (d) the temperature of the material is decreased
24. Fill in the blanks :
    (a) The resistance of a conductor ...... with increasing of its length.
    (b) The resistance of a conductor is ...... proportional to the area of cross-section of the conductor.

(c) The specific resistance is defined as the resistance offered by the two ...... faces of a unit cube.

(d) The equivalent resistance of the three elements connected in series is the ...... of all individual elements connected in series.

(e) The conductance ...... in parallel combination as more number of elements are added in parallel.

(f) The ...... of copper is more than the aluminium but less than the silver.

(g) The voltage drop in resistance is equal to ......

(h) The unit of electrical energy consumed is ......

(i) There is ...... path of current in series circuit.

(j) The resistance is measured in ......

25. To increase the resistance to corrosion ...... is added in steel.
(a) chromium (b) copper.
(c) aluminium (d) tungsten

26. The hardest neutral substance is ......
(a) diamond (b) cobalt
(c) platinum (d) uranium.

27. The order of resistivity for a conducting material is of the order of ......
(a) 10^{-8} to 10^{-6} Ω m (b) 10^{-4} to 10^{-2} Ω m
(c) 10^{-6} to 10^{-4} Ω m (d) 100 to 200 Ω m.

28. The order of resistivity for a semiconductor material is of the order of ......
(a) 10^{-8} to 10^{-6} Ω m (b) 10^{-4} to 10^{-2} Ω m
(c) 10^{-6} to 10^{-4} Ω m (d) None of the above.

29. The order of resistivity for an insulating medium is of the order of ......
(a) 10^{-8} to 10^{-6} Ω m (b) 10^{6} to 10^{7} Ω m
(c) 10^{3} to 10^{5} Ω m (d) 10^{12} to 10^{18} Ω m.

30. The total number of semiconductors in the periodic table so far are ......
(a) 2 (b) 8 (c) 6 (d) 13

31. The lightest particle of an atom ......
(a) neutron (b) proton (c) electron (d) nucleus.

32. The heavy particle of the atomic configuration is ......
(a) electron (b) proton
(c) free electron (d) None of the above.

33. If any electron is removed from an atom, it becomes ......
(a) positive ion (b) negative ion
(c) conductor (d) insulator

34. Mark the odd which is not the unit of power.
(a) amperes/volt (b) joules/sec
(c) amperes × volts (d) $(\text{amperes})^2 \times \text{ohms}$

35. The minimum possible charge that an ion can have is ......
(a) the charge of a neutron (b) the charge of an electron
(c) the total charge of electron and proton in the atom
(d) None of the above.

36. Two resistances when connected in series give the final resistance as 4 Ω and 1 Ω when connected in parallel. What are the values of each resistance?
(a) 2 Ω and 3 Ω (b) 2 Ω and 2 Ω.
(c) 4 Ω and 4 Ω (d) 1 Ω and 2 Ω.

37. Metals are good conductors of electricity because they have
(a) their valence orbit causes less attracting force between electron and nucleus.
(b) their valence orbit is completely filled.
(c) their valence orbit causes great force between free electrons and nucleus.
(d) None of the above.

38. ...... is not a conductor.
(a) Aluminium (b) Brass
(c) Copper (d) Mica

39. ...... is not an insulator.
(a) Aluminium (b) Asbustose
(c) Mica (d) Porcelain

40. The wire having 1.5 mm as diameter has cross-sectional area as ......
(a) 3 mm^2 (b) 1.79 mm^2 (c) 1.5 mm^2 (d) 6 mm^2

41. The electric pressure or potential difference that causes the insulation failure of the wire is called ......
(a) breakdown voltage (b) insulation resistance
(c) ampacity (d) None of the above.

42. The potential difference an insulation can withstand without breaking down is termed as ......
(a) insulation resistance (b) ampacity
(c) dielectric strength (d) None of the above.

43. The main purpose of soldering a joint is ......
(a) to increase its length (b) to increase its strength
(c) to increase its resistance (d) to make good electrical contacts.

CHAPTER 6

Sources of Electricity

INTRODUCTION

Q. **What do you mean by the source of electricity?**

Ans. The source by which we can get electricity or produce electricity, is called the source of electricity.

Q. **What are the different methods of producing electricity?**

Ans. These are, by means of friction, by chemical means, by mechanical means, by means of heat and by means of lightning effect etc.

Q. **What kind of electricity is produced by means of friction?**

Ans. The static electricity.

Q. **Give the example of producing electricity by means of chemical action?**

Ans. The cells and battery.

Q. **While the generators are producing electricity name the means adopted?**

Ans. By mechanical means.

Q. **How the thermal electricity is produced?**

Ans. It is because of heat when two dissimilar metals are heated up.

Q. **Give an example of producing electricity because of light?**

Ans. Photoelectric cell or solar cell.

EFFECTS OF ELECTRIC CURRENT

Q. **What are the effect of electric current?**

Ans. These are, heating effect, magnetic effect and chemical effect.

Q. **What is the heating effect.**

Ans. Whenever current is passed through a conductor, heat is produced in that conductor. This effect is called heating effect.

Q. **How this heat can be calculated?**

Ans. $H \propto I^2 \cdot R \cdot t$.
where H = heat, I = current, R = resistance and t = time.

Q. **What do you understand by the magnetic effect of current?**

Ans. Whenever the current is passed through a conductor, magnetic lines of force are produced around the conductor. This effect is called magnetic effect of current.

Q. **What is the direction of the current and the magnetic field thus produced?**

Ans. At right angles to each other.

Q. **The magnetic flux thus produced depends upon which factor?**

Ans. Current.

Q. **What is the chemical effect of current?**

Ans. It states that if the electric current is passed through an electrolyte it decomposes into its ions.

Q. **The electroplating depends upon which effect?**

Ans. Chemical effect of current.

Q. **What is an electric shock?**

Ans. It is nothing but sudden contraction of muscles with the feeling of pain, caused by electricity through the body of a person.

Q. **Actually speaking what is that factor which is responsible for electric shocks?**

Ans. Voltage which actually compels the current to flow.

Q. **What is the order of current with which a person feels a sensation bearable, painful, stop breathing or causes burning etc.?**

Ans. 1 mA to 8 mA, 8 mA to 15 mA and 20 mA and above, respectively. But it also depends upon the individual body resistance.

Q. **On which effect an incandescent lamp works?**

Ans. Heating effect of current.

Q. **How heat is developed?**

Ans. Because of the collision of electrons.

Q. **What is Joule's law of heating?**

Ans. The heat produced in a conductor is proportional to the square of current resistance and time of flowing current.

Q. **What are the different units of heat?**

Ans. Calorie, B.Th.U., joules.

Q. **What is one B.Th.U.?**

Ans. It is the amount of heat which is required to raise the temperature of one pound of water through 1°F.

Q. **How much calorie is equal to one joule?**

Ans. 1 cal = 4.2 joules.

Q. **How this heat is calculated?**

Ans. $H = 0.24 I^2 \cdot R \cdot t$ calories $= 0.24 \frac{E^2}{R} t = 0.24 E \cdot I \cdot t$ cal.

Q. **What do you mean by the thermal efficiency of the kettle?**

Ans. It is the ratio of output to input or say

$$\eta_{th} = \frac{\text{Heat useful}}{\text{Heat supplied}} \times 100$$

Q. **Can the efficiency be more than one?**

Ans. No. Always less than one.

Q. **Give the example of heating effect of current?**

Ans. Electric heater, electric oven, electric kettle, electric iron, soldering iron, etc.

OBJECTIVE TYPE MULTIPLE CHOICE QUESTIONS

1. The electric current is expressed in terms of ......
   (a) the rate of movement of electrons
   (b) the rate of movement of protons
   (c) the number of ions
   (d) the magnetic flux.
2. The e.m.f. induced in a conductor is directly proportional to the ......
   (a) rate of change of flux
   (b) rate of flow of electrons
   (c) magnitude of magnetic flux alone
   (d) the length of magnetic path.
3. The ratio of flow of an electric current is measured in ......
   (a) volts (b) ohms
   (c) farad (d) amperes
4. The e.m.f. induced in a conductor is measured in ......
   (a) volts (b) ohms
   (c) amperes (d) farads.
5. If two coulombs of electricity flows through a conductor in one seconds the current is ......
   (a) two amperes (b) one half amp

(c) four amp (d) one amp.

6. One coulomb of quantity is equal to ......
   (a) 0.386×10^{31} electrons (b) 7.61×10^{18} electrons
   (c) 6.24×10^{48} electrons (d) 624×10^{18} electrons.
7. The electric current can be realised by ......
   (a) electric shock (b) electric flashing
   (c) effects produced (d) electric heating.
8. The electric heater is the example of the ...... of electric current.
   (a) magnetic effect (b) chemical effect
   (c) heating effect (d) magnetic drag.
9. Whenever the electric current flows through an electrolyte it decomposes into ......
   (a) its atoms (b) its ions
   (c) its neutrons (d) its lattices.
10. The changes in the electrolyte when an electric current is passed are ...... changes.
    (a) magnetic (b) chemical
    (c) electrical (d) mechanical.
11. Whenever current flows through a conductor the heat is produced, this property is not so desirable in an electric ......
    (a) lamp (b) water heater
    (c) electric machine (d) soldering iron.
12. Whenever the heat is developed in a conductor it is governed by the ......
    (a) Ampere's law (b) Joule's law
    (c) Lenz's law (d) Newton's law.
13. The heat developed by an electric circuit is proportional to ......
    (a) $I^2 \cdot R \cdot t$ (b) $Z \cdot I \cdot t$
    (c) $m \cdot s \cdot t$ (d) $Q \cdot t$
14. Determine the rate of flow of coulombs through a 20 ampere hour battery if 20 ampere current is passed through it for 100 seconds.
    (a) 36 coulombs (b) 3.6 coulombs
    (c) 360 coulombs (d) 180 coulombs
15. One millivolt is equal to ......
    (a) 0.01 volt (b) 0.001 volt
    (c) 0.0001 volt (d) 0.1 volt
16. One kilo ohm is equal to ......
    (a) $10^3\ \Omega$ (b) $10^6\ \Omega$
    (c) $100^3\ \Omega$ (d) $10\ \Omega$

17. If the current is passed through a solution of $AgNO_3$ and it deposits the quantity of silver at the rate of ......, the electric current is said to be one international ampere.
    (a) 0.001118 gm/sec (b) 0.0118 gm/sec
    (c) 0.1118 gm/sec (d) 1.118 gm/sec.
18. One ampere represents the amount of electrons passing through a given point in a circuit in a specific ......
    (a) direction (b) duration
    (c) quantity (d) manner
19. The property of a substance which opposes the flow of current is known as ......
    (a) conductance (b) reluctance
    (c) resistance (d) admittance
20. The property of a circuit which helps the current to complete its path is known as ......
    (a) resistance (b) reluctance
    (c) retaintivity (d) conductance
21. The material through which current cannot flow is known as ......
    (a) liquid (b) solid
    (c) conductor (d) insulator
22. The ...... allows the current to complete its path easily.
    (a) resistance (b) insulator
    (c) conductor (d) semiconductor
23. The forbidden space is overlapped in ......
    (a) conductor (b) semiconductor
    (c) insulator (d) magnet
24. The ...... has more forbidden space than ......
    (a) conductor; insulator (b) insulator; semiconductor
    (c) insulator; conductor (d) semiconductor; insulator
25. The conductance through metals is due to the movement of ......
    (a) ions (b) neutrons
    (c) electrons (d) None of the above

CHAPTER 7

Fuse

INTRODUCTION

Q. **What is a fuse?**

Ans. Fuse is a safety device which protects the electrical circuit from taking overload current or protect the electrical circuit if more than pre-determined value of current flows.

Q. **On what principle does the fuse work?**

Ans. On the heating effect of current.

Q. **Why a fuse does not melt at normal current?**

Ans. Because at normal current, the amount of the heat developed is not sufficient to melt the fuse.

Q. **Which is the weakest point (electrically) in a domestic or industrial installation or circuit?**

Ans. Fuse.

Q. **What are the functions of a fuse?**

Ans.
1. To carry the normal working current safely without heating.
2. To protect the circuit in case of abnormal conditions of current.

Q. **What is the material used for fuse wire?**

Ans. It is lead in alloy tinned, copper wire, silver, gold, copper, aluminium, antimony etc.

Q. **Which material is generally used?**

Ans. Lead tin alloy and tinned copper.

Q. **Which size of lead tin alloy wire is suitable for 5 A current?**

Ans. 21 SWG wire.

Q. **Which tinned copper wire is suitable for 5 A, 15 A and 20 A?**

Ans. 35 SWG, 25 SWG and 23 SWG, respectively.

Q. **For precise work which type of fuse element is used?**
Ans. Silver.

Q. **Which material fuse is known as ideal fuse?**
Ans. Silver, because it is not subjected to oxidation.

Q. **When silver fuse is the ideal fuse, why it is not so commonly used?**
Ans. Because it is costly and cannot be afforded.

Q. **What do you understand by the fusing current?**
Ans. It is that minimum current at which the fuse element melts.

Q. **What is the current rating of fuse element?**
Ans. It is the safe value of current which a fuse element carries without heating or melting.

Q. **Name the factor which is the ratio of fusing current rating to the rating of fusing element?**
Ans. Fusing factor.

Q. **What is the percentage of lead and tin in the standard alloy fuse wire?**
Ans. 63% and 37% (lead, tin) respectively.

Q. **If a fuse wire for 10 A is required, which one will be suitable or used?**
Ans. Tinned copper.

Q. **Up to which capacity the lead tin alloy is used?**
Ans. Up to 5 A.

Q. **Is the fusing current constant or depends upon something else?**
Ans. It is not constant as it depends upon the position in which it is placed, its length and the atmospheric conditions.

Q. **If the two wires are used in twisted form what change in the fusing current will occur?**
Ans. The fusing current will be 1.7 times the fusing current of single wire.

TYPES OF FUSES

Q. **What are the types of fuses?**
Ans. These are :

1. Open type fuse
2. Semi-enclosed type fuse
3. Totally-enclosed type fuse

OPEN TYPE FUSE

Q. **What is the open type fuse?**
Ans. In this type of fuse, the fusing element is open to the atmosphere.

Q. **Where the open type fuses are used?**

Ans. Generally these are used on poles for giving supply to the consumers (for service mains).

Q. **What is the construction of the open type of fuse?**

Ans. In general construction it consists of a plate of porcelain having two terminals.

Q. **Where actually the fusing element is connected?**

Ans. Between these two terminals.

Q. **Is the fuse replacement easy in this type of fuse?**

Ans. No, the replacement is not so easy because one terminal is always energized.

Q. **As this element is open to the atmosphere so will the fusing current be constant?**

Ans. No, it also depends upon the surroundings.

Q. **Is there any possibility of fire with this type of the fuse?**

Ans. Yes, because when the element melts, sparking may fall on such articles which may catch fire then there is possibility of fire.

SEMI-ENCLOSED TYPE FUSE

Q. **What are the semi-enclosed type fuses?**

Ans. In this type of fuses, the fusing element is neither open to atmosphere nor enclosed.

Q. **Where the semi-enclosed type fuses are used?**

Ans. In domestic and industrial wiring.

Q. **What are the type of semi-enclosed type of fuses?**

Ans. Fuse cutout and kitket fuse.

Q. **What are the parts of fuse cutout?**

Ans. Two, base and cover (made of porcelain base).

Q. **Here the cover is used to cover the fuse element, why?**

Ans. Just to cover the element so that the sparking may not come out.

Q. **What are the parts of the kitket fuse?**

Ans. It is made of porcelain and has two parts—the base and fuse carrier.

Q. **How many terminals are there?**

Ans. Two.

Q. **Where the fuse element is placed?**

Ans. In the fuse carrier.

Q. **Which type of fuse is generally used in domestic installation?**

Ans. Kitket fuse.

Q. **What is the rating of kitket fuse generally used in house wiring?**
Ans. 15 A, 250 V.

Q. **Can the sparking fall outside in this type of fuse?**
Ans. No.

Q. **Can you replace the fuse wire easily?**
Ans. Yes, replacement of fuse element is easy.

Q. **How the fuses are rated?**
Ans. These are available according to the voltage and current rating.

Q. **According to voltage how these are classified?**
Ans. 250 V and 550 V gradings.

TOTALLY-ENCLOSED TYPE FUSE

Q. **What are the totally enclosed type fuses?**
Ans. In this type the fuse element is completely enclosed in a tube of glass or porcelain or fibre etc.

Q. **Is the fusing current of this type of fuses constant?**
Ans. Yes, it is constant because the fuse element is enclosed in a container or tube.

Q. **What are the types the totally enclosed fuses?**
Ans. HRC fuse and cartridge fuse.

Q. **What is the full form of the HRC fuse?**
Ans. High rupturing capacity fuse.

Q. **What is the construction of cartridge fuse?**
Ans. It has ceramic body sealed on both ends by metallic caps.

Q. **What is done to maintain the fusing current constant?**
Ans.
1. The fusing element is enclosed.
2. It is surrounded by silica sand or quartz powder to dissipate heat.

Q. **What is the indication of damaged fusing element of the HRC fuses?**
Ans. There is a white spot on the fuse body which becomes black when fuse melts out.

Q. **Where the totally enclosed type of fuses are used?**
Ans. In the places where high sensitivity and precise controlling is required.

Q. **Is the HRC fuse a rewirable fuse?**
Ans. No, and if at all any one does it, the accuracy will be certainly effected.

Q. **Can it be said that the enclosed type fuses are fuses with time lag arrangement?**
Ans. Yes, they actually delay some time, because the heat produced due to

dissipation does not melt the fuse immediately and hence delay some time.

Q. **Which fusing element is used in HRC fuses?**

Ans. Silver wire.

Q. **Is there any spark in the encloser and if it is, how it is quenched?**

Ans. The fuse melting may cause sparking or arc but the quartz powder which acts as an arc quenching medium quenches the spark.

Q. **Upto which capacity the HRC fuses can successfully deal?**

Ans. 46 kiloamperes.

Q. **Which fuses out of rewirable or HRC gives better overload protection?**

Ans. HRC fuse.

SOLDERING

Q. **What is soldering?**

Ans. Soldering is a process for joining two pieces of metal with the third metal generally known as solder.

Q. **What is a solder?**

Ans. It is an alloy used for soldering. Its melting point is very low.

Q. **What types of solder are available?**

Ans. Hard solder and soft solder.

Q. **Give the composition of hard and soft solders?**

Ans. In case of hard solders these are :

Spelter (copper + zinc)

Brazing (copper + zinc + antimony)

Silver solder (zinc + copper + silver)

In case of soft, it is an alloy of tin + lead (60 : 40 ratio).

Q. **What is the melting point of solder (60 : 40) used?**

Ans. 190°C.

Q. **Can you solder two metallic pieces with the solder which is having its melting temperature above the metals?**

Ans. Cannot, the melting temperature of the solder should be less the metal to be soldered.

Q. **What is flux?**

Ans. It is the agent which is used for soldering the joints.

Q. **What are the main contributions of the flux in jointing?**

Ans. It cleans the job and spreads the solder uniformly over the joint. It maintains the temperature of the solder and prevents the oxidation of the material.

Q. **What are the names of the flux generally used for solder?**

Ans. 1. Fluxite (Resin)

2. Aluminium chloride, killed acid and Eyre-7.

Q. **What is a solder?**

Ans. Solder is an alloy of lead and tin. It is used to join metals together by being melted.

Q. **What is desired about the solidification of the solder?**

Ans. The solder should solidify rapidly.

Q. **What is the composition of solder used for soldering for electrical work?**

Ans. Generally a cored solder is used. It has 60 : 40 (tin : lead) composition.

Q. **What do you mean by the cored solder?**

Ans. Cored solder means the solder has a core or cores of resin flux.

Q. **What is the utility of rapid solidification of solder as a matter of time?**

Ans. It performs the work quickly and shortens the time for the job.

Q. **Which types of soldering flux are used?**

Ans. Non-corrosive fluxes.

Q. **What are the types of non-corrosive fluxes?**

Ans. The generally used non-corrosive fluxes are of two types—resin fluxes and active fluxes.

Q. **What do you understand by resin fluxes?**

Ans. The resin flux is used to make the solder flow quickly over the place to be soldered. It also helps in transferring the heat. In addition to this it also helps to prevent the oxidising of the metal when heated.

Q. **Is there any additional benefit of using the active flux?**

Ans. Yes, it cleans the joint by removing grease and the oxidised film (if any) from the metal surface.

Q. **Are they non-corrosive at room temperature?**

Ans. Yes, at room temperature they are non-corrosive.

Q. **In which forms these are available in the market?**

Ans. Sometimes these are available in the form of paste or sometimes in liquid forms.

Q. **Which solder is advisable to use in electrical or electronics work?**

Ans. Cored solder.

Q. **Name the tool used for soldering purpose?**

Ans. Soldering iron.

Q. **How an electric soldering iron is heated?**

Ans. There is a heating element in the soldering iron. The supply is given to this element through flexible cord. The bit is heated as the electric current flows through the heating element; heat is produced which increases the temperature of the heating bit.

Q. **Which matter is used for bit of soldering iron?**

Ans. Generally copper.

Q. **In which wattage these are available?**

Ans. These are available in different wattage like 10 W, 15 W, 25 W, 30 W, 35 W, 40 W, 45 W, 65 W 125 W, 250 W etc.

Q. **Can you give some idea about the working voltage for it?**

Ans. It may be 220-230 V or less voltage for specially designed soldering irons.

Q. **Can you give some idea which factor actually decides the power rating of the soldering iron for that particular job?**

Ans. Yes, the quantity of heat and size to be soldered. The heat developed actually is a function of soldering iron wattage. There are certain places in which much heat will actually spoil the circuitry. There we need a soldering iron of low wattage for example in computer work, machines in video sets, VCR, etc. But there are certain other big places where this amount of heat is not of any use. There we can use the soldering iron of more wattage like 35 W, 45 W in radio, TV etc. For electrical work and jointing practice we use soldering iron of 65 W or 125 W, or sometimes for thimble practice also we prefer 250 W soldering iron.

Q. **Which care has to be considered for the selection of soldering iron?**

Ans. Select soldering iron with adequate power for the size of the work of soldering.

Q. **How a joint is soldered?**

Ans. Connect the soldering iron to the supply. Wait till it acquires the required temperature. Now attach the bit of the soldering iron to the job. As it reaches the fusion temperature of solder, apply solder on the joint or junction or point. Check the solder fillings or circulation around the junction and finally remove the iron. Let it cool.

Q. **How will you satisfy whether the required temperature is acquired or not?**

Ans. Let the tip be hot. As it starts giving a sensation of heat, apply the soldering iron on the tip. It should melt and should form a liquefy position of the solder.

Q. **Sometimes it is seen as we apply the solder on the tip, the solder becomes a round ball. What actually it indicates.**

Ans. The tip temperature is too high.

Q. **If it takes too much time to fuse the solder what actually it shows?**

Ans. The bit is yet to acquire more heat, the bit temperature is too low.

Q. **Is the cleaning of the junction to be soldered is important.**

Ans. The cleaning of junction or joint to be soldered is must, otherwise the solder will not get hold on the metal and joint will not be properly soldered.

Q. **If the tip temperature is very high can you do a fine soldering and if so what about the finish?**

Ans. The soldering will not be fine and moreover it is not advisable also. The finish of the job will carry a fat solder.

Q. **What could be the significance of the soldering at too high temperature?**

Ans. The solder will become a round ball, the metal may oxidise causing maladhesion of solder.

Q. **What precaution be taken to have a solder surface of a junction round and bright with metallic luster?**

Ans. As soon as the proper and required soldering is done, remove the soldering iron tip quickly because of surface tension of melted solder it will be round up around and will have a bright metallic lustre.

Q. **If the soldering iron tip is taken away too early what will happen?**

Ans. The solder will not solidify properly and flux adhesion may occur.

Q. **What will happen if the soldering iron is removed too late.**

Ans. The solder may spread too widely, the insulation may burn (if insulated cable is used), the part or object may be damaged, serious oxidation may result spoiling the goods than proper soldering.

Q. **Name the factors which decide the quality of good finish of soldering?**

Ans. The appreciable and beautiful finish of the soldered joint depends on cleanliness of the parts or junction to be soldered, temperature control of the soldering tip and the skill of soldering.

Q. **What are the methods of jointing two metallic pieces?**

Ans. Soldering, brazing and welding.

Q. **What is brazing?**

Ans. Soldering at high temperature is known as brazing. In this brass is generally used as solder so it is called brazing.

Q. **What is welding?**

Ans. Welding is the process of joining two metallic pieces with the help of third, known as electrode.

Q. **How much voltage is generally used for welding?**

Ans. It is done at low voltage and high current, without load the voltage is 40-60 V and drops down to 20-25 V and average current having the range from 80-400 A and even more.

Q. **What are the tests to be performed for testing the heating appliance?**

Ans. 1. Continuity test.
2. Earth fault or leakage test.

Q. **What do you mean by the continuity test?**

Ans. This test is performed to test the continuity of the element.

CABLE TERMINATION

All the apparatuses or equipments working on electricity are to be supplied and for that the connections are made. The wire termination is must. All terminations must be made to provide better electrical continuity and should be in such a way as not to come in contact with other metallic parts and cables. Loose connections or gap in termination will cause overheating. The over-heating of cable and terminals may result in fire. The long termination or uninsulated wire or extended wire may result in shock, further deterioration of insulation and even short-circuit. The touching of unwrapped conductors of a stranded cable may cause short-circuit, thus it goes without saying that wrong termination of cable will lead to overheating, damaging insulation, short-circuit i.e. electrical hazards.

Q. **What are the over comes of wrong terminations?**

Ans. The wrong termination may lead to overheating of termination point of the cable, short circuit or earth leakage etc.

Q. **What is the main object of good termination?**

Ans. Good termination means to provide better electrical conductive path or better electrical continuity.

Q. **How crimping is done?**

Ans. In this crimping is done with the crimping tool. The conductor is inserted into the crimping terminal and then crimped.

Q. **What important precaution must be observed in crimping?**

Ans. The very important precaution which should be taken is that we should select a crimp terminal that matches the conductor diameter and the dimensions of the connecting screw terminal.

Q. **Which portion is called crimping?**

Ans. The cable ends connected to terminating lug is called crimping.

Q. **On compression type connectors which crimping tool is used?**

Ans. Suitable compression tool that is also a type of crimping tool.

Q. **What is the main aim of the pressure.**

Ans. It is to reduce the electrical resistance between the cable and terminal. Otherwise the more resistance will produce more heat resulting in overheating and damage.

Q. **How this tool is operated?**

Ans. It is otherwise called a hand crimping pliers. Hold the fitting into the jaws and crimp by squeezing the handles.

Q. **What is the importance of indenting?**

Ans. The properly executed crimp will indent the top of the lug and the indentation will grip the conductor effectively.

Q. **Where the crimping on the barrel is preferred?**

Ans. In the centre of the barrel.

Q. **If the crimping is done too deep, is it advisable?**

Ans. The deep crimping will not reduce contact resistance but it will increase it. So a properly crimping tool be selected.

Q. **What is the general size of terminal lug crimping hand tools?**

Ans. The terminal lug crimping pliers are available in lengths ranging from 18 cm to 30 cm.

Q. **Can you give some idea about any of the crimping pliers specifications regarding the wire to be crimped.**

Ans. The crimping tool used for 26 to 12 SWG wires. The other may be for other specifications.

Q. **Can you use different head and jaws with one pliers?**

Ans. Yes, at a time only one head and jaws will be used but both can be replaced for another size of wire.

Q. **Which factor, the crimping plier or crimping jaws decides the shape of the crimp?**

Ans. The shape of the jaws actually decides the shape of crimping.

Q. **What can be the general probability of usual accident with this tool?**

Ans. Trapping a finger as the operating cycle is non-reversible.

Q. **What do you means by the non-reversible operating cycle?**

Ans. Once the jaws are pressed or squeezed these can be released only be applying further pressure.

Q. **What will you note while selecting the mechanical size?**

Ans. The mechanical size that is considered as tongue size and thickness, hole size and shape etc.

Q. **What electrical consideration is taken into account for selecting the crimping terminal?**

Ans. Current-carrying capacity, which will actually decide the mechanical dimensions like cross-sectional area and the wall thickness of the terminal.

Q. **Generally which material is used for the material?**

Ans. Copper or brass but now-a-days aluminium, nickel and steels are also used but not so promptly.

Q. **Which shapes of these crimping terminals are in frequent use?**

Ans. Rectangular, ring, spade and flanged spade shapes are in general use.

OBJECTIVE TYPE MULTIPLE CHOICE QUESTIONS

1. The fuse which is used on the poles is known as ......
   (a) open type fuse (b) closed fuse
   (c) HRC fuse (d) cutout
2. The fuse melts because of ......
   (a) low current (b) heavy current
   (c) no current (d) rated current
3. The fuse should always be connected on ......
   (a) earth wire (b) phase wire
   (c) neutral wire (d) None of the above
4. The characteristics of the fusing element should be a wire of ......
   (a) low resistance (b) high resistance
   (c) high melting point
   (d) low resistance and low melting point.
5. A fuse wire melts because of ......
   (a) the heating effect of current
   (b) the magnetic effect of current
   (c) the chemical effect of current
   (d) the phenomenon of producing X-rays.
6. The function of the fuse in electrical circuit is to protect ......
   (a) the insulation resistance of the cable
   (b) overloading of the circuit
   (c) overvoltage
   (d) insulation resistance and excess voltage.
7. The length of the cartridge fuse depends upon the ...... of the circuit.
   (a) power (b) current
   (c) voltage (d) type of electricity
8. The current rating of the fuse is determined by considering ......
   (a) wattage of the individual lamp
   (b) capacity of the smallest wire in the circuit
   (c) voltage of the circuit
   (d) rating of the largest wire of the circuit.
9. For better performance the ...... is suitable.
   (a) open type fuse (b) kitket fuse

(c) HRC fuse (d) cutouts

10. In HRC fuse, the HRC stands for ......
    (a) high rupturing capacity fuse
    (b) high resistance capacity fuse
    (c) heavy resistive carbon fuse
    (d) heavy reactive capacitance fuse.

11. ...... is the process of joining two metals with the help of third nobel metal at low temperature.
    (a) Brazing (b) Soldering
    (c) Welding (d) Annealing.

12. The melting point of (60 : 40) solder is ......
    (a) 190°C (b) 1000°C
    (c) 210°C (d) 760°C

CHAPTER 8

Resistance

INTRODUCTION

Q. **What is resistance?**

Ans. It is the property of the substance which opposes the flow of current through it.

Q. **What is the unit of resistance?**

Ans. Ohm.

Q. **How it is measured and name the instruments with which it is measured directly?**

Ans. It can be measured by ammeter-voltmeter method, by comparison method, by Wheatstone bridge (meter bridges, post office box) and directly by means of ohm-meter.

Q. **What is AVO meter?**

Ans. It is also known as multimeter and can be used to measure current, voltage and resistance in amps, volts and ohms respectively.

Q. **How it is represented?**

Ans. By letter 'R' and unit is ohm.

Q. **What is one ohm?**

Ans. If a potential difference of one volt is applied to any resistance and it allows one ampere current to flow then the value of the resistance is one ohm.

Q. **What are the other units of resistance?**

Ans. These are micro ohm = $10^{-6}\ \Omega$

milli ohm = $10^{-3}\ \Omega$

mega ohm = $10^{6}\ \Omega$

kilo ohm = $10^{3}\ \Omega$

Q. **The property which opposes the flow of current is known as resistance, what will you call the property which helps the current to complete its path?**

Ans. Conductance.

Q. **What is the unit conductance?**

Ans. Mho.

Q. **On what factor does the resistance of a conductor depend?**

Ans. It depends upon the length, area of cross-section, material and temperature.

Q. **What is the relationship between resistance and length of the conductor?**

Ans. $R \propto l$. It is directly proportional.

Q. **The resistance of the 5 metre length of wire is 10 Ω, what will be the resistance of the same wire of 10 metres?**

Ans. 20 Ω.

Q. **If you decrease the length of the resistance wire, what will be the effect on the resistance of this wire?**

Ans. It will decrease.

Q. **Is the resistance of a conductor inversely proportional to the area of cross-section?**

Ans. Yes.

Q. **If the area is made double, what will happen to the resistance?**

Ans. It will be reduced to half.

Q. **If the length is increased twice of a certain wire and area reduced to half what will be the resistance?**

Ans. Four times increase in the resistance.

Q. **How the resistance of the conductor is concerned with the material?**

Ans. The resistance of a conductor is directly proportional to the nature of the material i.e., $R \propto \rho$.

Q. **What should be the property of a good resistance?**

Ans. The temperature coefficient should be low, secondly it should be able to withstand the temperature.

Q. **If you increase the temperature of the pure metals how it will effect the resistance of that material?**

Ans. It will increase regularly depending upon the temperature.

Q. **If an insulator is heated up, what will happen to its resistance?**

Ans. It will decrease.

Q. **How the resistance of an electrolyte is affected by the increase in temperature?**

Ans. It will decrease.

Q. **What is the formula for calculating the resistance if the length, area of cross-section and specific resistance of a conductor are given?**

Ans. $R = \frac{\rho \cdot l}{a} \, \Omega.$

SPECIFIC RESISTANCE

Q. **What is the specific resistance?**

Ans. It is the resistance offered by the two opposite faces of a unit cube or can otherwise be defined as the resistance offered by a conductor having unit length and unit area of cross-section.

Q. **With which letter it is represented?**

Ans. ρ.

Q. **What is the unit of specific resistance?**

Ans. It micro ohm cm or micro-ohm-m.

Q. **What is the specific resistance of copper, silver and graphite?**

Ans. 1.731 $\mu\Omega$ cm, 1.56 $\mu\Omega$ cm and 3000 $\mu\Omega$ cm, respectively.

Q. **Which is the best conductor?**

Ans. Silver.

Q. **What is the difference between resistivity and specific resistance?**

Ans. Both are same.

TEMPERATURE COEFFICIENT

Q. **What do you understand by the temperature coefficient?**

Ans. It is the ratio of the increase in resistance for 1°C rise in temperature to actual resistance at 0°C. Or it can otherwise be defined, the increase in resistance per ohm per degree is known as temperature coefficient of that material.

Q. **With which letter it is denoted?**

Ans. α.

Q. **What is the unit of temperature coefficient?**

Ans. It is the ratio and ratio has no unit, but given at certain degree temperature.

Q. **What is the temperature coefficient of copper?**

Ans. $\alpha = 0.0043$.

Q. **If a certain resistance has R_0 resistance at 0°C and after heating to t°C the resistance changes of R_t Ω, what will be the temperature coefficient of the material?**

Ans. $\alpha = \frac{R_t - R_0}{R_0 \cdot t}$

Q. **If the temperature coefficient is α and R_0 resistance at 0°, what will be the resistance at t°?**

Ans. $R_t = R_0(1 + \alpha t)$.

Q. **Can you find out the resistance of a given specimen if the resistance at 0° is not given?**

Ans. Yes.

$$\alpha = \frac{R_{t_2} - R_{t_1}}{R_{t_1}(t_2 - t_1)}$$

Substituting the known values we can find out the resistance.

Q. **If two dissimilar metals having different temperature coefficients are heated up, will these expand to same length?**

Ans. No, the expansion will be according to their temperature coefficients.

Q. **Where this principle is applied?**

Ans. In the thermal or bimetallic relays.

CIRCUIT

Q. **What is an electric circuit?**

Ans. The complete path taken by the electric current is called electric circuit.

Q. **What is an ideal circuit?**

Ans. The electric circuit in which current completes its path through a controlling device, consuming device and safety device, is known as ideal circuit.

Q. **What are the different types of circuits?**

Ans. These are closed circuit, open circuit and short circuit.

Q. **What is an open circuit?**

Ans. The circuit in which live wire or return wire or resistance is broken and current cannot flow or complete its path, is called as open circuit.

Q. **What is short circuit?**

Ans. The circuit in which live or positive wire and negative or neutral wire meets without the consuming device, is called short circuit.

Q. **What is the ideal circuit or closed circuit?**

Ans. The circuit in which the current completes its path through the controlling, consuming and safety devices, is known as the closed circuit or in short if the negative and positive or phase and neutral completes its path through a consuming device then it is known as the ideal circuit or closed circuit.

Q. **A simple circuit consists of which elements?**

Ans. The simple or closed circuit consists of a voltage source, a safety device (fuse), a controlling device (switch) and a consuming device.

Q. **What is the necessity of a safety device?**

Ans. For the purpose of safety, if by chance the circuit develops a short circuit defect, the battery or the voltage source will supply excessive current which may damage wires, accessories or source, so for all these safety the safety device is introduced.

Q. **If a generator is running and the load switch is opened, which condition is or under which type of circuitry this will be considered?**

Ans. Open circuit.

Q. **If in a wall socket no load is connected, under which type of circuit will it come?**

Ans. Open circuit.

Q. **In this how much will be the resistance, if measured, between the terminals.**

Ans. Infinitely high resistance.

Q. **If the phase and neutral or positive and negative an joined without load, under which category this circuit will come?**

Ans. Short circuit.

Q. **What will be the condition of load current in this case?**

Ans. The current supplied by the battery or generator will be extremely high.

Q. **Which element in this condition helps the circuit to save?**

Ans. Fuse, the safety device.

Q. **If the fuse is not provided in the circuit what will be the situation?**

Ans. It will be the worst possible; because the source will supply high and extremely high current until it is discharged completely. In case of generator the winding will be damaged or the prime mover will get stopped.

Q. **How the fuse will protect the circuit?**

Ans. The fuse, because of high current will melt out and result an open circuit; as a result there will not be any current and circuit will be saved.

Q. **Will the rate of flow of electrons be same with different wattage lamp when supplied by the same voltage?**

Ans. No, they will be different depending upon the wattage of the lamp.

Q. **In which case it will be more?**

Ans. It will be more with more wattage lamp and less with less wattage lamp.

Q. **Name the factor which can be determined by the counts of electrons through a cross-section of the conductor?**

Ans. Electric current.

Q. **What is the duration generally taken to mark the current as one amp or two amp etc.?**

Ans. Actually the number of electrons passing though a point in one second are termed as one amp or two amps.

Q. **How does a short circuit behave in terms of resistance?**

Ans. A short circuit behaves like a path of very low resistance as compared to the normal circuit resistance.

Q. **What are the types of short circuits?**

Ans. A short circuit may be of two types—the partial short circuit and full or dead short circuit.

Q. **In case of partial short circuit actually what happens?**

Ans. The circuit current increases beyond the predetermined value. It may be due to faulty appliances, damaged insulation or leakage etc.

Q. **What defects may results due to short circuit?**

Ans. Excess current, damaging of wires and cable, circuit components, power source, possibility of fire due to intense heat generated in the conductors.

Q. **How the damage of the possibility of short circuit be prevented?**

Ans. By using proper fuses, using circuit-breakers.

Q. **How the fuse and CB will be connected?**

Ans. They all will be in series with the load.

Q. **Can you detect a short circuit by using a voltmeter?**

Ans. Yes, if we connect a voltmeter across the resistance or load where the possibility of short circuit is there, the reading will be zero volt which certifies the short circuit.

Q. **How an open circuit be indicated by a voltmeter?**

Ans. If you connect a voltmeter across a resistance or load when there is a possibility of open circuit, the voltmeter will read full line voltage.

Q. **If you connect one ammeter in series with an open-circuited load, can you tell the reading obtained?**

Ans. Yes, it will be zero ampere. The ammeter will not deflect at all.

Q. **Where are the possibilities of probable open circuit?**

Ans. The open circuit may be in the terminals and contacts of switches, circuit fuse may be burnt out, in the connection wire, and damaged load or resistance.

Q. **If there are three resistances connected in series and you are measuring voltage across individual resistances, can you check or identify the open-circuited resistance just by using a voltmeter?**

Ans. Yes, it can be easily identified. The resistances where the voltmeter reads zero voltage are not defective but if across any individuals the voltmeter reads full voltage it means that particular resistance is defective.

Q. **If in a circuitry the voltmeter does not give any voltage with an individual resistance can you tell the reason?**

Ans. There is a possibility that any of the two resistances are open-circuited.

Q. **Can you detect the open circuit by the use of ohmmeter?**

Ans. Yes, it is also an effective instrument for locating the open circuit. Connect the leads (keeping the selector switch in ohms suitable range) and then see the readings at the place (load or resistance) where the open circuit is there. The meter will read an infinity resistance.

Q. **If in a series connections (like ladi of small bulbs) one has to locate open circuit. How will you proceed?**

Ans. Connect one lead to the line, other end left free, check the phase by the other wire or line tester and proceed step by step. The place of open circuit will be automatically indicated by not glowing the test lamp or tester (take 15 W lamp for lines testing).

COMBINATION OF RESISTANCES

Q. **What are the different methods for connecting the resistances?**

Ans. These are series, parallel and compound or series parallel.

SERIES COMBINATION

Q. **What is the series circuit?**

Ans. The circuit in which the resistances are connected in the shape of a chain is called series circuit. In this the end of one resistance is connected with the starting of the second and the end of second with the starting of third, the connections are known as series connections.

Q. **Give any example of series connections?**

Ans. The decoration light (ladi) having a number of bulbs (6 V, 12 V) connected across 220 V mains.

Q. **How many paths of current are possible in series connections?**

Ans. Only one.

Q. **Will the current be divided according to their resistances?**

Ans. No, it will be the same in all the resistances connected in series.

Q. **How the voltage drop across the individual is calculated?**

Ans. It can be calculated by multiplying the current and the resistance of the respective element connected.

Q. **Will the sum of the drop be equal to the voltage given?**

Ans. Yes.

Q. **How you can calculate the equivalent resistance if there are r_1, r_2 and r_3 resistances connected in series?**

Ans. $R_{eq} = r_1 + r_2 + r_3\ \Omega$.

Q. **If you increase the resistance connected in series will the conductance be decreased or increased?**

Ans. Decreased.

Q. **How the power consumed in series circuit is calculated?**

Ans. $I^2 \times R$ where I is the current drawn and R is the equivalent resistance.

Q. **What will be the current if equivalent resistance is R and voltage given is V?**

Ans. It is $\frac{V}{R}$ amp.

Q. **(a) If 2 Ω, 3 Ω and 5 Ω resistances are connected in series across 10 volts, what will be the total resistance and current drawn?**

Ans. 10 Ω and 1 amp.

Q. **(b) What will be the voltage drop across 3 Ω resistance if current is 1 A?**

Ans. 3 V.

Q. **If you add more resistances in series will the current increase?**

Ans. Decrease.

Q. **If two lamps of 60 W, 250 V and 100 W, 250 V are connected in series across 250 V, 50 c/s mains, which lamp will glow more?**

Ans. 60 W lamp.

Q. **Why this bulb will give more light?**

Ans. Because the voltage drop across this lamp will be more.

Q. **Which lamp has more resistance?**

Ans. The 60 W lamp has more. 60 W lamp resistance is 1041 Ω and 100 W lamp resistance is 625 Ω.

Q. **When both lamps carry equal current why don't they illuminate to same level?**

Ans. Because the 60 W lamp is designed for low current than 100 W lamp. Secondly the voltage across lamp is also more than 100 W lamp.

Q. **What will be the total power in this case?**

Ans. It will be less than the element of less wattage connected in series in the circuit, i.e. 37.5 W.

Q. **Why the combined power is less?**

Ans. Because in series the reciprocal of equivalent power is the sum of reciprocals of individual powers. So the equivalent power will be only 37.5 W.

Q. **Out of this power consumption which lamp will consume more power?**

Ans. 60 W lamp.

Q. **Why this bulb consumes more power?**

Ans. Because the current is same in both lamps. Their resistances are different so it is very much obvious that the lamp having more resistance will consume more power ($W = I^2R$), and the resistance of 60 W lamp is more than 100 W lamp.

Q. **When both are in series can we have such an arrangement so that both lamps may give same light as these are designed/rated for?**

Ans. Yes we can do that.

Q. **How?**

Ans. If we could arrange double the voltage and place a suitable resistance across 60 W lamp to have equivalent resistance of 60 W combination equal to 625 Ω, then same voltage will be dropped across each lamp resulting normal voltage and normal light.

Q. **If two lamps of 25 W and 40 W, 110 V are connected in series across 220 V, 50 c/s mains, can you tell which lamp will glow more?**

Ans. 25 W lamp.

Q. **Can you estimate the voltage across this lamp?**

Ans. It will be more than 110 V.

Q. **How will you arrange/which arrangement would you suggest so that both lamps may give their rated output on 220 V?**

Ans. When both are in series as the resistances are different, the 25 W lamp has more resistance than the 40 W lamp. Now we have to connect a resistance of suitable value in parallel to the 25 W lamp so that the combined resistance of 25 W lamp circuit would be equal to 40 W lamp. Thus the voltage drop across each lamp will be half i.e. the rated voltage operating the lamp at their respective ratings.

Q. **If in this unbalanced load, different wattage lamps are connected in series across the double voltage, which lamp would have the chance to fuse?**

Ans. Less wattage lamp.

Q. **If you are going into a workshop for testing and removing fault, which testing lamp would you prefer two lamps of same wattage connected in series or single lamp of 250 V?**

Ans. Two lamps of same wattage connected in series. This system is called dual lamp testing holder.

Q. **A 100 W lamp is connected in series with a 1000 W heater. Now if**

the 100 W lamp is replaced with 200 W lamp, what will happen to the output of the heater?

Ans. The heater output will increase.

Q. **Why it will increase?**

Ans. The 100 W lamp has more resistance than the 200 W lamp so the voltage across heater will increase, increasing the output of the heater. The 200 W lamp is capable of carrying more current than 100 W lamp.

Q. **If a resistance of 5 Ω having wire size 0.01 mm in diameter and another resistance of 25 Ω having wire size as 0.025 mm are connected in series which resistor will dissipate more power if connected in series across 220 V mains.**

Ans. 25 Ω resistor.

Q. **When a heater is connected on mains, it glows and gives heat which could be easily realised. Now the heater lead is also in series with the heater, why this lead does not glow or starts dissipating heat as that of heater because same circuit is flowing through it also?**

Ans. The same current is undoubtedly flowing through the lead but it has less resistance and the power which is a product of I^2R will be considerably less thus it does not appear to glow.

Q. **If 2 A, 250 Ω resistance is connected in series with 3.5 A, 180 Ω resistance, the safe current circuit of the series circuit will be how much?**

Ans. 2 A.

PARALLEL COMBINATION

Q. **What is a parallel circuit?**

Ans. The circuit in which the resistances are so connected that their starting are connected to one terminal and ends to the another terminal that it is known as parallel circuit.

Q. **Name the function which is constant to all elements?**

Ans. Voltage.

Q. **What will be the current in individual resistance?**

Ans. Voltage divided by the individual resistance.

Q. **What will be the equivalent resistance in parallel circuit?**

Ans. It is $\frac{1}{R_{eq}} = \frac{1}{r_1} + \frac{1}{r_2} + \frac{1}{r_3} + \ldots$

Q. **Will the total resistance increase or decrease with the addition of resistances in the circuit?**

Ans. It will decrease.

Q. **What will be the conductance, if you add more components in the circuit?**

Ans. It will increase with the addition of components.

Q. **How will you calculate the total current drawn by the circuit?**

Ans. It can be by two ways. First find out the total resistance of the circuit and current will be voltage divided by this resistance.

Secondly, find out the individual branch currents and add them that will be the current drawn by the circuit.

Q. **If you add more elements will the total current increase or decrease?**

Ans. Increase.

Q. **All the appliances connected in your home are in series or in parallel?**

Ans. In parallel.

Q. **If these be connected in series, what will happen to the light of the bulbs?**

Ans. These will give dim light.

Q. **If 60 W, 220 V and 160 W, 220 V lamp is connected in parallel which lamp will drop more voltage?**

Ans. The voltage in parallel circuit is same so same 220 V will be across each lamp.

Q. **Which lamp out of 60 W, 220 V and 100 W, 220 V will have more resistance?**

Ans. 60 W lamp.

Q. **When both lamps are in parallel what will be their equivalent power consumed?**

Ans. It will be 160 W.

Q. **All the lamps or appliances in domestic use are connected in which fashion?**

Ans. All are in parallel.

Q. **What happens if all are connected in series?**

Ans. Every individual will not get required voltage and will not work satisfactorily as in series connections the algebraic sum of the voltage drops will be equal to the voltage fed.

SERIES-PARALLEL COMBINATION

Q. **What are the series parallel connections?**

Ans. The connections in which the number of elements are connected in series as well as in parallel are known as compound or mixed connections or series-parallel connections.

Q. **How you will calculate the value of current and voltage in this circuit?**

Ans. That will be turn by turn solving parallel combinations and series combinations etc.

OHM'S LAW

Q. **Name the terms in which the relationship is proved by Ohm?**

Ans. Current and voltage.

Q. **State the Ohm's law?**

Ans. Ohm's law states that in any closed circuit the current is directly proportional to the voltage, provided the physical conditions of the circuit are kept constant.

i.e. $I \propto V$

Q. **The ratio of voltage and current is equal to a constant. Name the constant?**

Ans. Resistance.

Q. **What is the relationship between the voltage, current and resistance?**

Ans. $\frac{\text{Voltage}}{\text{Current}} = \text{Resistance}$

Q. **How the current is related with the resistance?**

Ans. Current is inversely proportional to the resistance.

Q. **If the resistance of a certain appliance is increased will the current increases or decrease?**

Ans. Decrease.

Q. **If a circuit having 20 Ω resistance and is connected across 20 V supply, what is the current?**

Ans. One ampere.

Q. **What do mean by the physical conditions of the resistance?**

Ans. The length, area of cross-section, and temperature etc.

Q. **If in any circuit the voltage is constant and some value of current is flowing without disturbing the voltage can you change the current flowing?**

Ans. Yes, by changing the resistance.

Q. **State the formula for calculating, current, voltage and resistance?**

Ans.

$$\text{Voltage} = \text{Current} \times \text{Resistance}$$

$$\text{Current} = \frac{\text{Voltage}}{\text{Resistance}}$$

$$\text{Resistance} = \frac{\text{Voltage}}{\text{Current}}$$

KIRCHHOFF'S LAW

Q. **Which type of circuits are solved by this law?**

Ans. The circuits which are neither series circuit nor parallel circuit and not even series and parallel circuits, but the complex circuits.

Q. **How many laws are there in Kirchhoff's law?**

Ans. These are two—current law or point law and voltage law or mesh law.

CURRENT LAW

Q. **State current law?**

Ans. It states that the sum of the currents meeting at a point is equal to the sum of the currents going away from that point.

Or

The algebraic sum of the currents meeting at a point is equal to zero.

Q. **Can any current be left at any junction in an electric circuit?**

Ans. No.

VOLTAGE LAW

Q. **What is voltage law?**

Ans. According to this law in any closed circuit or mesh the algebraic sum of the products of the current and resistance of each part of the circuit is equal to the resultant voltage fed to the circuit.

Or

The algebraic sum of the IR drops and EMFs fed to the closed mesh is always zero.

Q. **How the sign of drops or rise of potential difference is marked?**

Ans. Whenever the rise in potential difference is there the drops are marked as positive and fall in potential is marked as negative (on the other hand some books prefer the rise in potential as negative and fall as positive).

Q. **What precautions should be taken strictly into consideration while marking?**

Ans. In this case choose only and only one procedure for marking and it should be strictly followed.

OBJECTIVE TYPE MULTIPLE CHOICE QUESTIONS

1. Ohm's law states that the current in any closed circuit is proportional to the voltage, provided the ...... conditions remain constant.

   (a) thermal (b) physical

(c) electrical (d) mechanical

2. The current varies in ...... proportional to the resistance in any D.C. closed circuit.
   (a) direct (b) inverse
   (c) semi-direct (d) linear
3. A lamp of 25 watts is connected in series with 40 watts lamp, 25 watts lamp glows more because its resistance is ......
   (a) in series with 40 watts lamp
   (b) not connected in parallel
   (c) more than 40 watts lamp
   (d) less than 40 watts lamp.
4. The current taken by a load when connected to some voltage is ......
   (a) voltage × resistance (b) voltage ÷ resistance
   (c) wattage ÷ voltage (d) wattage ÷ resistance
5. The power taken by a load can be calculated by V^2/R, $V \cdot I$ and ......
   (a) $I^2 \times R$ (b) $I^2/5 \times R$
   (c) $I^2 \log R$ (d) $R \log (I)^2$
6. The resistance of 100 watts, 250 V lamp is ......
   (a) 25000 Ω (b) 625 Ω
   (c) 0.4 Ω (d) 2.5 Ω
7. What will be the ampere meter reading if 2 volt is applied to 10 Ω resistance?
   (a) 20 amperes (b) 2 amperes
   (c) 0.2 ampere (d) 4 amperes.
8. A 100 Ω resistance is connected across 300 volts the power taken is ......
   (a) 9000 watts (b) 30000 watts
   (c) one watt (d) 900 watts.
9. The resistance of an appliance is 10 Ω. What additional resistance must be placed in series to limit the current to 3 amperes at 24 V.
   (a) 30 Ω (b) 80 Ω
   (c) 70 Ω (d) 90 Ω
10. The three different resistance are connected in such a way that the current in all is same, so these are connected in ......
    (a) series (b) parallel
    (c) series-parallel (d) complex.
11. The current drawn by a series circuit ...... with the increasing of the resistance in the circuit.
    (a) remains the same (b) decreases
    (c) increases (d) reaches to zero.

12. If two elements of 1000 watts are connected in series, the total wattage will be ......

(a) 1000 watts
(b) 2000 watts
(c) 500 watts
(d) 4000 watts

13. The voltage remains the same across each apparatus, the combination is connected in ......

(a) series
(b) parallel
(c) series-parallel
(d) complex.

14. A resistance of 20 Ω is connected to 20 V supply the current is ......

(a) one ampere
(b) 0.5 ampere
(c) 2 amperes
(d) 20 amperes

15. The sum of the currents meeting at a point is equal to the sum of currents going away from that point is the ...... law.

(a) current law
(b) voltage law
(c) Coulomb's law
(d) Joule's law.

16. The law used to determine the force between two current-carrying conductors is ......

(a) Joule's law
(b) Lenz's law
(c) Biot Savart's law
(d) Faraday's law

17. Brass is an alloy of copper and zinc in the ratio of ......

(a) 66% : 34%
(b) 50% : 50%
(c) 34% : 66%
(d) 80% : 20%

18. Which resistance material is used for precision work?

(a) Manganin
(b) Carbon
(c) Eureka
(d) Constantan

19. Nichrome is an alloy of ......

(a) nickel, chromium
(b) nickel, iron
(c) chromium
(d) copper, zinc.

20. The resistivity of iron and nickel is of the order of ......

(a) 10^{-9} Ωm
(b) 10^{-7} Ωm
(c) 10^{-5} Ωm
(d) 10^{3} Ωm.

21. Arrange gold, aluminium, silver and copper in their order of increasing order of resistivity.

(a) Ag, Cu, Au, Al
(b) Cu, Ag, Al, Au
(c) Ag, Al, Au, Cu
(d) Ag, Au, Al, Cu.

22. Among Kanthal, Manganin, Nichrome and Tantenum, which has highest resistivity?

(a) Manganin
(b) Kanthal
(c) Tantenum
(d) Nichrome.

23. Which has the lowest temperature coefficient ......
    (a) Au
    (b) Cu
    (c) Ag
    (d) Al
24. Density of gold is approximately equal to the density of ......
    (a) Tantenum
    (b) Tungsten
    (c) Platinum
    (d) Nichrome.
25. Generally ...... are used for conductors.
    (a) copper, aluminium
    (b) copper, silver
    (c) aluminium, silver
    (d) copper only.

CHAPTER 9

Capacitor

INTRODUCTION

Q. **What is a capacitor or condenser?**

Ans. It is a static device which stores the electricity in the shape of charge.

Q. **What is charge?**

Ans. Electricity at rest.

Q. **What are negatively charged particles?**

Ans. The particles which have more number of electrons than protons are called negatively charged.

Q. **If the number of protons are more than the electrons what will be the charge of the body?**

Ans. Positive.

Q. **In which unit the capacity of a condenser is measured?**

Ans. Farad. Other units are microfarad and picofarad etc.

Q. **What is micro and picofarad?**

Ans. 10^{-6} farad = 1 μF and 10^{-12} farad = 1 μμF or picofarad.

Q. **On what factors the capacity of a condenser depends?**

Ans. It depends upon the area of the plates, dielectric used and the thickness of the dielectric used.

Q. **What is the relationship between the capacitance of a condenser and the area of the plate?**

Ans. $C \propto A$ (directly proportional).

Q. **If the thickness of the dielectric is increased will the capacitance be changed and if yes, how?**

Ans. Yes, $C \propto \frac{1}{t}$

Q. **If the temperature of the dielectric is increased what will happen to the dielectric strength?**

Ans. It will be changed and certainly in the reciprocal order.

Q. **What is the dielectric strength?**

Ans. It is represented in voltage which it can withstand without puncturing, generally it is in kV/mm.

Q. **Which dielectric will be better, having less dielectric strength or more dielectric strength?**

Ans. More.

Q. **How does the moisture affects the dielectric strength?**

Ans. Decreasing with the increasing of moisture.

Q. **Write the formula for calculating the capacitance if area of cross-section of plate, dielectric thickness etc. are given.**

Ans. $C = \frac{KA}{4\pi t}$

Q. **What is a multiplate condenser?**

Ans. The condenser which has more than two plates is known as multiplate condenser.

Q. **Which plates are more, the positive or negative?**

Ans. Positive.

Q. **What are the properties of charge?**

Ans.
1. Like charges repel each other.
2. Unlike charges attract each other.

Q. **What is coulomb?**

Ans. It is the unit of charge.

Q. **What is the Coulomb's law?**

Ans. The force of attraction or repulsion between two charges is proportional to the product of their charges and inversely proportional to the square of distance between them.

Q. **What is the mathematical representation of Coulomb's law?**

Ans. $F = \frac{\lambda \times q_1 \times q_2}{d^2}$

where λ is a constant.

Q. **Is there a field around a charged body?**

Ans. Yes.

Q. **Name the field?**

Ans. Electric field or electrostatic field.

Q. **What is an electric field?**

Ans. It is the region in which the electric charge experiences some force or some stress.

Q. **How the stress around the field is experienced?**

Ans. Through imaginary lines of force.

Q. **Who invented the lines of force for electrostatic field?**

Ans. Faraday.

Q. **How do the lines of force travel?**

Ans. From positive terminal to negative terminal.

Q. **If a charged body is brought near the uncharged body what will happen?**

Ans. It will be charged having opposite polarity.

Q. **What the phenomenon of inducing opposite charge to a uncharged body is called?**

Ans. Electrostatic induction.

Q. **What is electric flux density?**

Ans. It is the lines per unit area.

Q. **What is one coulomb?**

Ans. A condenser has a charge of one coulomb when one ampere of current flows for one second in that condenser.

One coulomb = one amp/sec

Q. **What is the capacitance?**

Ans. It is the rate of charge per voltage i.e. $E = Q/V$.

Q. **What do you mean by charging of a condenser?**

Ans. Whenever the condenser takes current from a source of supply, it is called as charging.

Q. **What is meant by discharging?**

Ans. Whenever the charged condenser is connected across the load (resistance) and it delivers the stored electricity, it is called as discharging.

Q. **Name the substance which is placed in between the plates of a condenser?**

Ans. Dielectric.

Q. **What is the permittivity or specific inductive capacitance?**

Ans. It is the ratio of capacitances of the same condenser with the given material as dielectric to the capacitance of the same condenser with air as dielectric.

Q. **What is the symbol generally used for the dielectric constant?**

Ans. K.

Q. **What is the unit of dielectric constant?**

Ans. It is ratio, so it is only a number.

Q. **What is dielectric strength?**

Ans. It is the voltage at which any dielectric is punctured or can be said as the maximum voltage which a dielectric can withstand without puncturing.

Q. **What is the general unit in which the dielectric strength is shown?**

Ans. kV/mil or kV/mm.

Q. **What are the factors upon which the dielectric strength depends?**

Ans. The thickness of the dielectric, temperature and moisture also,

Q. **What is the relationship between the dielectric strength and the temperature of the dielectric?**

Ans. Dielectric strength $\propto \dfrac{1}{\text{Temperature}}$

Q. **What is the dielectric constant of mica?**

Ans. 5.8.

Q. **What is the dielectric strength of mica?**

Ans. 40-156 kV/mil.

Q. **What is the dielectric constant of air?**

Ans. One.

Q. **What are the factors which affect the capacity of a condenser?**

Ans.
1. Area of the plates.
2. Dielectric constant.
3. Thickness of the dielectric.

COMBINATION OF CAPACITORS

Q. **How many groups are possible?**

Ans. Series, parallel and mixed or series-parallel connections.

SERIES CONNECTION

Q. **What is series connection?**

Ans. In this type one end of one condenser is connected with the start of th second condenser and the end of the second with the start of thir condenser thus forming a chain. This type of connections are known a series connection.

Q. **How the voltage in series connection are divided?**

Ans. The sum of the voltage drops across each condenser is equal to th voltage fed to the group.

Q. **If three capacitors having C_1, C_2 and C_3 capacities are connected i series how much will be the capacity of the group?**

Ans. $\frac{1}{C_{eq}} = \frac{1}{C_1} + \frac{1}{C_2} + \frac{1}{C_3}$

Q. **What will be the charge on each capacitor?**

Ans. It will be same and equal to $C \cdot V$.

Q. **Why the charge is same?**

Ans. Because there is only one path of current to flow.

Q. **Can you calculate the voltage across each capacitor?**

Ans. Yes. $v_1 = \frac{q}{c_1}$

where v_1 is the voltage across any capacitor, q is the charge and c_1 is the capacity of that particular capacitor.

PARALLEL CONNECTION

Q. **What is parallel connection?**

Ans. If the starting of all condensers are connected to one terminal and endings of the condensers are connected to another terminal, then the connections are called parallel connection.

Q. **Will the charge be common to all?**

Ans. No.

Q. **Name the function which is common to all condensers?**

Ans. Voltage.

Q. **What will be the total charge of the combination?**

Ans. $Q = q_1 + q_2 + q_3 + \ldots$

where Q = equivalent charge, q_1, q_2, and q_3 are the individual charges.

Q. **The charge acquired by each capacitor depends upon the voltage or capacity or both?**

Ans. Both.

Q. **What is the equivalent charge in case of parallel circuit if the equivalent capacity is C?**

Ans. $Q = C \times V$.

Q. **What is the formula for finding out the total equivalent capacity of the combination?**

Ans. $C = c_1 + c_2 + c_3$.

where C = equivalent capacity; c_1, c_2 and c_3 are the individual capacities.

SERIES-PARALLEL CONNECTIONS

Q. **What is the series-parallel connection?**

Ans. The combination in which the capacitors are connected in series and in parallel both.

Q. **How the charge and other functions are calculated in series-parallel circuit?**

Ans. Step by step by solving series and parallel combinations separately.

Q. **What are the types of the condensers according to the values?**

Ans. Fixed and variable.

Q. **Classify the condensers according to their dielectrics?**

Ans. Paper, mica, electrolytic and air condensers.

Q. **What is the main base for the types of the condenser?**

Ans. Dielectric used.

Q. **What dielectric is used in paper condenser?**

Ans. Paper.

Q. **Name the dielectric used in air condensers?**

Ans. Air.

Q. **What do you understand by the variable capacitor?**

Ans. The capacitor by which a variable capacitance is obtained, is known as variable capacitor.

Q. **What are the different ways of obtaining variable capacitance?**

Ans.
1. By using a number of condensers.
2. By using variable air condenser.

Q. **Where variable condensers are used?**

Ans. Generally in radio works and wherever variable capacitance is required.

Q. **What do you mean by the fixed condenser?**

Ans. The condensers which are having fixed values.

Q. **Name the part of the paper condenser?**

Ans. Mainly these are parafined paper as dielectric, two thin foils and rolled in a cylindrical form.

Q. **What are the approximate ranges of the paper condenser?**

Ans. It is from 0.0001 μF to 0.01 μF in different capacities.

Q. **What are the approximate ranges of the mica condenser?**

Ans. 50 pF to 500 pF.
(pF = picofarad)

Q. **What are the uses of mica condenser?**

Ans. These are used in radio work and telecommunication work etc.

Q. **What are the types of electrolytic condensers?**

Ans. Wet type and semi-dry type.

Q. **Are these available for low capacity or high capacity?**

Ans. High capacity.

Q. **How the condensers are tested?**

Ans. By means of the screw driver, by means of high resistance meter and by means of measuring the capacity of the capacitor.

Q. **How a condenser is tested by means of the screw driver?**

Ans. In this test, capacitor is charged and then it is short circuited by means of a screw driver. If it gives spark, it is alright and if it does not give any spark it is open circuited. If the fuse blows off as soon as it is connected for the charging then it is short circuited.

Q. **What should be the reading of the voltmeter if the condenser is short circuited?**

Ans. Meter will give continuous reading.

Q. **If the meter gives reading at the time of switching on to D.C. what about the condition of the condenser while high resistance testing is adopted?**

Ans. Condenser/capacitor is alright.

Q. **What do you understand by a cylindrical condenser?**

Ans. A cylindrical condenser consists of two co-axial cylinders of different radius, separated by the dielectric.

Q. **Give an example of a cylindrical capacitor?**

Ans. A single core cable.

Q. **What is the energy stored by a condenser?**

Ans. It can be given by $0.5CV^2$ where C is the capacity, V the voltage where it is connected.

Q. **What are the other formulae for calculating the energy stored?**

Ans. $0.5Q \cdot V, 0.5\frac{Q^2}{C}$ joules.

Q. **What is the time constant of a circuit having capacitor?**

Ans. The time constant of a circuit is the time during which the charging current falls to 0.37 of its initial maximum value.

Q. **If you connect one ammeter in series with the capacitor, will it give any reading (D.C. circuit)?**

Ans. Yes, at the time of starting.

Q. **Why not after some time?**

Ans. Because in steady state condition the capacitor behaves like an insulator for D.C. circuit.

Q. **If the capacitor is connected to A.C. through an ammeter, will it give any reading?**

Ans. Yes, continuously.

Q. **What is the permittivity of the free space?**

Ans. 8.854×10^{-12} farad/metre.

Q. **What is the electric force?**

Ans. The electric force or electric intensity at any point in the field or space around the charge is measured by the force which will act on a unit positive charge placed at that point.

OBJECTIVE TYPE MULTIPLE CHOICE QUESTIONS

1. To calculate the quantity of energy stored by a capacitor ......
   (a) multiply the voltage across the capacitor terminals by its half capacitance.
   (b) multiply the voltage across the capacitor terminals by its half resistance.
   (c) divide the voltage across the capacitor terminal by its half resistance.
   (d) divide the voltage across the capacitor terminals by its half capacitance.
2. To calculate the time period of a sine wave ......
   (a) multiply its peak voltage by 2.83,
   (b) multiply its frequency by 2.83,
   (c) divide 1 by its RMS voltage,
   (d) divide 1 by its frequency.
3. The capacity of a condenser is proportional to ......
   (a) area of its plates
   (b) volume of its plate
   (c) the specific resistance of the plate material
   (d) the temperature coefficient of the plate material.
4. The capacity of a capacitor is inversely proportional to ......
   (a) the temperature of the dielectric
   (b) the material of the dielectric
   (c) the thickness of the dielectric
   (d) the permeability of the material and inductance.
5. The capacitors in series have the same ......
   (a) voltage (b) capacity
   (c) charge (d) energy loss.
6. The equivalent capacity of the combination in series ...... with the increasing of the capacitors in series.

(a) increases (b) decreases
(c) remains the same (d) becomes zero.

7. The capacitors are named according to the ...... used.
(a) material of the plate (b) dielectric used
(c) enclosures used (d) voltage

8. The capacity of the capacitor is given by ......
(a) $1/2CV^2$ (b) voltage and current
(c) $VI \cos \phi$ (d) I^2R.

9. The time constant of an R-C circuit is given by ......
(a) L/R (b) CR
(c) CR^2 (d) $1/2Q/V$.

10. The bank of the capacitors is used to ...... the power factor of the A.C. line.
(a) decrease (b) increase
(c) meet the voltage fluctuations
(d) improve the copper losses.

11. The unit of the capacity of a capacitor is ......
(a) microampere (b) milliohm
(c) henry (d) farad

12. The capacitors are used to take ......
(a) kW (b) kVA
(c) kVAR (d) kVL.

13. If an ammeter and a capacitor are connected in series on D.C. circuit, the ammeter will ......
(a) give continuous reading
(b) initially and then will come to zero
(c) not give any reading
(d) give only half the reading.

14. The capacitor stores the electricity in the shape of ......
(a) dynamic charge (b) static charge
(c) current electricity (d) molecules.

15. The capacitance can be obtained by using ......
(a) chokes (b) reactors
(c) capacitors (d) resistors.

16. To increase the capacitance of a capacitor, the distance between the plates must be ......
(a) closer to each other (b) made smaller in size
(c) further apart (d) made short circuited

17. The equivalent capacity of two capacitors in series combination is given as ......

(a) $C_{eq} = \frac{C_1 + C_2}{C_1 - C_2}$ (b) $C_{eq} = \frac{C_1 \cdot C_2}{C_1 - C_2}$

(c) $C_{eq} = C_1 + C_2$ (d) $C_{eq} = \frac{C_1 - C_2}{C_1 + C_2}$

18. In case of parallel combination the equivalent capacity of two capacitors is given as ......

(a) $C_{eq} = C_1 + C_2$ (b) $C_{eq} = \frac{C_1 + C_2}{C_1 - C_2}$

(c) $C_{eq} = \frac{C_1 \times C_2}{C_1 + C_2}$ (d) $C_{eq} = \frac{C_1 - C_2}{C_1 + C_2}$

19. The energy stored in an electrostatic field of a capacitor is given as ......

(a) $1/2CV^2$ joules (b) $1/2QV^2$ joules

(c) $0.5CI^2$ joules (d) $0.5MV^2$ joules

20. If a charged capacitor is short circuited then a ...... will produce.

(a) spark (b) nothing

(c) voltage (d) torque.

21. The unit of permittivity is ......

(a) N/m (b) farad/m

(c) Wb/m (d) farad/m^2

22. If the distance between the plates of a condenser is decreased to half distance, the capacity of the condenser will ......

(a) increase four times (b) decrease four times

(c) increase twice (d) decrease to half

23. The field intensity is measured in ......

(a) coulomb (b) coul/m

(c) N/coul (d) Wb/m

24. The static electric charge can be removed by ......

(a) grounding its frame

(b) increasing surrounding moisture

(c) providing special grounding around the appliance

(d) None of the above.

25. Electrolytic capacitors are generally manufactured in ...... capacity.

(a) constant (b) low

(c) high (d) variable

26. If four capacitors of 16 μF each are connected in series their resultant capacity will be ......

(a) 64 μF (b) 8 μF
(c) 32 μF (d) 4 μF

27. If four capacitors 16 μF each are connected in parallel, their resultant capacity will be ......

(a) `64 μF (b) 8 μF
(c) 32 μF (d) 4 μF

28. One farad is defined as ......

(a) one coulomb/amp (b) one coulomb/volt
(c) one joule/sec (d) None of the above.

29. The capacitive reactance of 100 μF capacitor is 31.83 Ω at 50 c/s. What will be its capacitance at 100 c/s?

(a) 31.83 Ω (b) 15.92 Ω
(c) 63.66 Ω (d) 100 Ω

30. A 100 μF pure capacitor will dissipate at 250 V, 50 c/s ......

(a) 100 watts (b) zero watt
(c) 200 watts (d) 250 watts.

CHAPTER **10**

Wires, Cables, Wiring and Earthing

CONDUCTOR AND INSULATOR

Q. **What do you understand by conductor and insulator?**

Ans. The substance or material through which electricity can pass easily, is known as conductor and the substance through which electricity cannot pass easily or offers high resistance in the flow of current is known as the insulator.

Q. **What are the properties of a good conductor?**

Ans. It should have low specific resistance, good mechanical strength, ductile, malleable easily jointable, cheap and free from external atmospheric effects.

Q. **How the conductivity and resistivity are related?**

Ans. Inversely proportional to each other.

Q. **Which is the best conductor?**

Ans. Silver.

Q. **What is the resistivity of silver?**

Ans. It is 1.64 Ω/m.

Q. **Name five conductors according to their conductivity?**

Ans. Copper, aluminium, tungsten, zinc and brass.

Q. **What is the conductivity of aluminium in comparison to copper?**

Ans. 60%.

Q. **Why aluminium is widely used nowadays?**

Ans. Because it is available in India in plenty and is cheaper than copper.

Q. **Name the conductor which is generally used in OH line?**

Ans. ACSR (Aluminium Conductor Steel Reinforced).

Q. **Where brass and lead are used?**

Ans. Brass is generally used for making the terminals of switches, holders and other accessories. Lead is used for soldering, filling in cables, jointing, and making protective covering of underground cables etc.

Q. **What should be the properties of a good insulator?**

Ans. High ohmic resistance, high specific resistance, good mechanical strength, good dielectric strength, free from atmospheric changes.

Q. **Which is the best insulator?**

Ans. Dry air.

Q. **Name some other insulators?**

Ans. Vulcanised Indian rubber, mica, ebonite, leatheroid, insulating oil, porcelain, asbustose, shellac and slate etc.

Q. **What is the dielectric strength of vulcanised rubber, mica and ebonite?**

Ans. 30-50 kV/mm, 20-60 kV/mm, 30-40 kV/mm, respectively.

Q. **What are the properties of mica and where it is generally used?**

Ans. It is transparent, fireproof, and does not moist. It is good conductor of heat. It is used in soldering iron, electric kettle, commutator and so many other places.

Q. **What do you know about leatheroid?**

Ans. Leatheroid is a good insulating material. It is affected by grease, oil etc. It burns at high temperature. It is used for slot insulation and coil insulation for the winding work also.

Q. **Name any ten insulators?**

Ans. Porcelain, asbustose, empire cloth, pressphan paper, fiber, cotton tape, black adhesive tape, bitumen compound and gutta-percha.

WIRES AND CABLES

Q. **Name the tool generally used for finding out proper diameter and gauge of the conductor?**

Ans. The micrometer and standard wire gauge.

Q. **How the various types of cables used in house wiring are classified?**

Ans. The wires or cables used in house wiring may be classified according to :

1. The conductor,
2. Voltage grading,
3. Number of cores, and
4. Type of insulation used.

Q. **What are the classifications according to the cores and type of insulation?**

Ans. According to cores these are :

1. Single core,
2. Two cores and
3. Three cores

and according to the insulation used, these are :

PVC wire, CTS or TRS wire, lead covered wire, weatherproof wire and flexible wires etc.

Q. **What is VIR wire?**

Ans. It is the Vulcanised Indian Rubber wire.

Q. **What is the construction of VIR wire?**

Ans. It is the vulcanised Indian rubber wire. It has one or more conductors of aluminium or copper which are covered with rubber insulation and then cotton tape and cotton braiding and compound.

Q. **What do you understand by TRS wire, PVC wire and CTS wire?**

Ans. TRS wire—Tough Rubber Sheathed wire

PVC wire—Poly Vinyl Chloride wire

CTS wire—Cab Tough Sheathed wire.

Q. **What protective covering is made over the lead covered wire?**

Ans. Lead sheathed covering.

Q. **What are the different parts of the weatherproof wire?**

Ans. It has copper or aluminium conductor with rubber insulation, then cotton tape, then PVC sheathed and finally cotton braiding with sindur etc. for weatherproof covering.

Q. **What are the advantages of using multistrand conductors than single solid conductor?**

Ans. The multistrand conductors are more flexible, durable having more heat dissipating area, and the skin effect is also reduced to considerable extent than that of single solid conductor.

Q. **How the cables are specified according to the sizes?**

Ans. The cables are specified as 1/18, 3/22, 3/20 7/20 SWG etc. which indicates that the numerator shows the conductors in the cable and denominator shows the gauge of the conductors used. Nowadays aluminium conductors are used which are specified as 1.5 mm^2, 2.5 mm^2, 4 mm^2, 6 mm^2 and so on.

Q. **On what factors the size of the conductor depends?**

Ans. The current capacity of the load, voltage drop and mechanical strength also.

WIRING

Q. **What is wiring?**

Ans. Wiring means the systematic layout of the wires or to connect the electrical load to supply in a proper manner and according to Indian Electricity Rules.

Q. **What are the types of wiring?**

Ans. Two types—the domestic wiring and industrial wiring.

Q. **What are the types of house or domestic wiring?**

Ans. Cleat wiring, Casing and Capping wiring, Batten wiring, Conduit pipe wiring and Lead covered wiring.

Q. **What are the types of conduit wiring?**

Ans. 1. Surface conduit pipe wiring.
2. Concealed conduit or underground wiring.

Q. **What are the different systems of distribution?**

Ans. Tree system and distribution board system.

Q. **What are the main points to be considered for wiring for particular place?**

Ans. Durability, safety, cost, appearance, and accessibility.

CLEAT WIRING

Q. **What are the different materials used in cleat wiring?**

Ans. Porcelain cleats (two ways, three ways), VIR wire or PVC wire, screws, blocks, and boards and different electrical accessories.

Q. **What are the advantages of cleat wiring?**

Ans. It requires less labour and workmanship, less time, less cost, extension and removal is also easy.

Q. **What are the disadvantages of the cleat wiring?**

Ans. Less life, less efficiency, dust and dirt spoil the insulation, no safety from fire, smoke, sunlight, rains, dampness and mechanical damages etc.

Q. **How much should be the distance between two cleats?**

Ans. It should not exceed more than 60 cm and safe distance is 30 cm.

Q. **Is it a permanent types of wiring?**

Ans. No, it is a temporary wiring.

Q. **What is the recommended period for this type of wiring?**

Ans. It should not be used for more than six months, but if it is required to be used for more than six months then it should be further tested and checked properly.

CASING AND CAPPING WIRING

Q. **What do you understand by casing and capping wiring?**

Ans. In this type of wiring the VIR wire runs in the grooves of teak wood casing and casing is covered with capping, therefore this wiring is known as casing and capping wiring.

Q. **What sizes of casing and capping are used?**

Ans. 1½″ × ⅝″, 2″ × ¾″, 3″ × 1″, 1¾″ × ⅝″, 2½″ × ¾″, 3½″ × 1¼″ and 4″ × 1¼″ etc.

Q. **What are the advantages of this wiring?**

Ans. It has good appearance, long life, wires are safe from mechanical disturbances. If it is painted and round cleats are provided then it is free from dampness and vermins also.

Q. **What are the disadvantages of this wiring?**

Ans. The wiring requires more skilled labour and skilled workmanship, more cost, no safety from fire, extension is not so easy, fault location is not so easy.

Q. **Can you run two opposite polarity wires in the same groove?**

Ans. No.

Q. **How the wiring can be safe from dampness?**

Ans. By providing round porcelain cleats below the casing and by painting the casing and capping.

Q. **Can it be buried under the wall?**

Ans. No, it should never be done.

Q. **Where this wiring is used?**

Ans. In houses but should not be done in damp places, where the risk of fire and fumes exists.

Q. **How much should be the distance between the gitties or cleats?**

Ans. The maximum distance is 3′, for casing and capping up to 2″ size but for more than 2″ it should be kept 2′ maximum.

Q. **If the wiring is to be taken in other rooms (crossing the wall) how it should be taken?**

Ans. In that case wires should be taken through a conduit pipe or PVC pipe.

Q. **How much should be the distance between the screws on capping?**

Ans. It should not be more than 6″.

Q. **In 2″ × ¾″ casing capping how many wires should be run of 3/22 SWG?**

Ans. Four maximum.

Q. **Name the different joints used in casing and capping wiring?**

Ans. Straight joint, right angle joint, corner joint, tee joint, bridge tee joint, cross joint and cross bridge joint etc.

Q. **Where the straight joint, corner joint and right angle joint are used?**

Ans. In domestic wiring these joints are used, straight joint where the length of casing and capping is to be increased; the corner joint at the wall corners and right angle joint where the wiring is to be bend at 90°.

Q. **Why bridge joints are preferred?**

Ans. Because the wires of opposite polarity should not cross each other.

BATTEN WIRING

Q. **What is a batten wiring?**

Ans. The wiring in which wires are run on the wooden batten fixed in clips.

Q. **What are the different materials used in batten wiring?**

Ans. Teak wood batten, round block, boards bends and corners, clips, nails, wires, wiring accessories and round cleats.

Q. **What are the different sizes of batten?**

Ans. ½″ × ½″, ¾″ × ½″, 1″ × ½″, 1¼″ × ½″, 1½″ × ½″ and 2″ × ½″ etc.

Q. **What are the advantages of the batten wiring?**

Ans. Good appearance, cheap, protection against the moisture if painted and round porcelain cleats are used, semiskilled worker can do it, repair and extension is easy and easily accessible.

Q. **What are the disadvantages of this wiring?**

Ans. No safety from fire, and shock, cannot be done in open space because atmospheric effects will damage the insulation of wires.

Q. **What should be the distance between the clips?**

Ans. It should be in horizontal line not more than 4″ and in vertical line should not be more than 6″.

Q. **What are the types of clips?**

Ans. Joint clips, link clips.

Q. **Which type of wires are mostly used in batten wiring?**

Ans. CTS or TRS, PVC and weatherproof wires.

Q. **What should be the bend of wire to protect the insulation safety of wire?**

Ans. It should not be less then six times the external diameter of the wire. The bend should not be sharp.

Q. **What should be the maximum distance between the gitties in batten wiring?**

Ans. It should not exceed more than 2.5 m.

Q. **If the wiring is to be crossed through the wall what precaution is to be observed?**

Ans. The batten should never be buried under the wall, a conduit pipe should be used to cross the wall. The bushings should also be provided on both sides of conduit pipe.

Q. **Where the batten wiring is used?**

Ans. The wiring is done even in moist places, generally it is used in house wiring.

Q. **How the earth wire is taken or wired?**

Ans. Generally it is drawn below all wires on the batten and fixed in the same clips.

Q. **If lead covered wire is used on the batten what do you call this wiring?**

Ans. Lead covered batten wiring.

Q. **Why the lead covered wire is used?**

Ans. It can protect the conductors/wire from dampness, atmospheric changes, corrosion, and mechanical injury etc.

Q. **Why earth wire is also run along the LCC wire?**

Ans. The lead covering also serves the purpose of earth continuity conductor but for better earthing, and to keep the potential of earth wire at zero potential, it is run along the wire.

Q. **How many cores LCC wires are available?**

Ans. Single core, twin cores.

Q. **What are the disadvantages of LCC wiring?**

Ans. Costly, skilled worker and workmanship is required, extension and changes are very difficult.

Q. **How much should be the curvature in LCC wiring?**

Ans. It should not be less then nine times the external diameter of the LCC wire used.

Q. **Where the LCC wiring is used?**

Ans. In telephones, in service mains of domestic installation, up to meter etc.

CONDUIT PIPE WIRING

Q. **What are the types of conduit pipe?**

Ans. Light gauge and heavy gauge.

Q. **How the light gauge and heavy gauge conduit pipes are prepared?**

Ans. **Light gauge** pipes are prepared from sheets and these are thin, so threading cannot be done, these are not suitable for good wiring.

Heavy gauge pipes are either solid drawn or welded and prepared from mild sheets, threading can be done.

Q. **How many types of wiring is done with conduit pipe?**

Ans. Surface conduit pipe wiring and concealed or underground conduit pipe wiring.

Q. **What is the difference between the surface and concealed conduit pipe wiring?**

Ans. In surface conduit wiring the conduit pipe is run above the surface of wall or ceiling etc. fixed with saddles. In concealed wiring the conduit pipe is buried under the wall or ceiling.

Q. **Name the different accessories used in conduit pipe wiring?**

Ans. Conduit pipe, conduit couplings, conduit elbow, conduit junction boxes (single way, two ways, three ways, four ways), conduit tee. conduit check nuts, conduit saddles, conduit nipples, conduit bushings, reducers, flexible conduit pipe, rigid and flexible couplings etc.

Q. **What are the common sizes of conduit pipes?**

Ans. ½″, ¾″, 1″, 1¼″, 1½″, 2″, 2½″ etc.

Q. **How the threadings are done on the pipe?**

Ans. By the conduit die set.

Q. **What accessories are used for fixing the conduit pipe?**

Ans. Saddles, clips, hooks, crampets and girder clips.

Q. **What are the different methods of drawing the wires in the conduit pipe?**

Ans. By drawing system and by pushing method.

Q. **What are the advantages of conduit pipe wiring?**

Ans. Wires are safe from mechanical damage, extension and repair easy, alternation is easy, fault location not difficult, wiring has more life and it is shock-proof if earthing is done properly.

Q. **How earth wire is drawn in case of surface and concealed conduit wiring?**

Ans. In surface conduit outside the pipe, clipped with earth clips and in concealed conduit wiring the earth wire is drawn through the pipe.

Q. **How many wires of 1/2 SWG and 3/22 SWG can run through the conduit pipe of ¾″?**

Ans. For 250 V grading—six wires both 1/18 and 3/22 SWG wires; for 650 V grading—four wires both 1/18 and 3/22 SWG.

Q. **Which type of wiring is suitable for blacksmith shop, wet places, and cold storage?**

Ans. Conduit pipe wiring, batten with cleats CTS or PVC wiring.

MEGGER TESTING

Q. **What is a megger?**

Ans. Megger is an instrument used for measuring the insulation resistance of the wiring.

Q. **What is the necessity of testing?**

Ans. Before energizing a new installation, the testing is an essential factor to avoid any possibility of accident because of supply, short circuit and leakage current etc.

Q. **What maximum leakage current is permissible?**

Ans. The leakage current should in no case exceed 1/5000th of the total load current.

Q. **What should be the insulation resistance of the new insulation?**

Ans. The insulation resistance of the new installation should not be less than $\frac{50}{\text{Number of outlets}}$ MΩ.

Q. **Can you calculate the leakage current of the installation/appliance, if the voltage and insulation resistance are given?**

Ans. $\text{Leakage current} = \frac{\text{Voltage}}{\text{Insulation resistance in M}\Omega}$

Q. **How many terminals are there in a megger?**

Ans. Line and earth—only two.

Q. **What should be the speed of the handle while the instrument is in operation?**

Ans. 160 rpm generally or as per the instructions supplied by the manufacturer.

Q. **What are the different tests to be performed for testing a new installation?**

Ans.
1. Continuity test,
2. Insulation test between conductors of opposite polarity.
3. Insulation resistance between conductors and earth.
4. Polarity test.
5. Earth continuity test.

Q. **Can you use the megger if the installation is already energized?**

Ans. No.

Q. **What actually is tested in case of continuity test?**

Ans. The continuity of the wires throughout the installation.

Q. **Will the switches be in off or on position?**

Ans. For effective testing, the switches are 'ON' turn by turn.

Q. **What will be the position of the holders and plug points in case of continuity test?**

Ans. The lamps are inserted in holders and the sockets are either short circuited or the appliances are connected,

Q. **What will be the reading if all things are alright?**

Ans. Zero with every observation.

Q. **What is the reading in case of the insulation resistance test between the conductors of opposite polarity?**

Ans. Infinity.

Q. **Should we place the lamps in holders in case of insulation resistance testing?**

Ans. No.

Q. **If you don't 'ON' the switches will it make any difference?**

Ans. Yes, the half wire insulation resistance will not be tested.

Q. **In case of insulation resistance test between conductor and earth, where the terminals of megger should be connected?**

Ans. One terminal on earth (E-terminal) and after twisting both positive and negative or phase and neutral both are connected to another terminal marked L.

Q. **What should be the insulation resistance reading of the megger?**

Ans. Infinity or the meter should read at least one mega ohm.

Q. **What do you test in polarity test?**

Ans. Whether the phase wire is connected to the switches or not.

Q. **Where the phase wire should be connected?**

Ans. To the switches.

Q. **How the polarity test is performed?**

Ans. One terminal of the megger is connected with phase and the other terminal turn by turn to the switches. The megger should give zero reading and if not there is defect. Check and remove it.

Q. **How the earth continuity is tested?**

Ans. Connect one line terminal to E (earth) at the main board termination of earth point, and then connect the other terminal of megger with earth point of three-pin plug sockets, metallic body of ceiling fan, brass batten holder and bracket etc. (where earthing is done), the megger should read zero.

Q. **If there are more number of circuits, then should it be tested with each circuit or only one circuit testing is sufficient?**

Ans. It should be tested with each circuit independently.

Q. **If the insulation resistance of a new installation is 7.5 MΩ, can it be energized or not?**

Ans. Yes it can be energized.

Q. **A single way switch is rated 5 A, 250 V and a fuse of 5 A, will it be advisable to use a single way switch in place of fuse?**

Ans. No, fuse is a safety device and a switch is a controlling device, secondly a switch will not protect the line, if the load current is more than 5 A.

Q. **What are the common faults in wiring?**

Ans. Leakage, open circuit, short circuit and reverse connections.

EARTHING

Q. **What do you mean by earthing?**

Ans. Earthing means the connections of a metallic body with the general mass of earth for the purpose of safety from shock.

Q. **What is the necessity of earthing?**

Ans. The main purpose is the safety from shock.

Q. **Which wire is connected to earth (i) two wire D.C. system (ii) three wire D.C. system. (iii) A.C. 3 ϕ alternator?**

Ans. (i) In case of D.C. two wire system the negative wire is connected to earth.

(ii) In case of D.C. three wire system the middle wire is connected to earth.

(iii) In case of three phase generators, the star point or neutral is connected to earth.

Q. **What is an earth electrode?**

Ans. It is a metal pipe or plate which is connected electrically with the general mass of earth and is buried under the earth.

Q. **What do you understand by earthing lead?**

Ans. The wire which is connected with the earth electrode is known as earthing lead.

Q. **What is earth continuity conductor?**

Ans. The conductor which is connected to the earthing lead and which connects all the metallic parts which are to be earthed without breaking its continuity, is known as earth continuity conductor.

Q. **What are the different methods of earthing?**

Ans. Pipe earthing and plate earthing.

Q. **What is the main importance of earthing?**

Ans. Safety against electric shock.

Q. **In case of earthing what is the role of earth?**

Ans. Earth plays an important role. It behaves like a short-circuited thick conductor.

Q. **What about the potential difference of the earth in that case?**

Ans. Earth is supposed to be at zero potential.

Q. **What is the purpose of earthing?**

Ans. The purpose of earthing is to provide protection to human beings, equipment, and circuit against electric shock (dangerous or excessive voltage).

Q. **Which part or portion of the apparatus/equipment machine/wiring be earthed?**

Ans. All the metallic portions of the system except current-carrying conductors.

Q. **What do you understand by system earthing?**

Ans. The earthing of supply system i.e. earthing of neutral wire of supply system.

Q. **What is the utility of this system earthing?**

Ans. It limits and provides a balance voltage system. For example if neutral is properly earthed, the voltage in 3 ϕ across one phase and neutral will be established and balanced (balanced load).

Q. **What is equipment earthing?**

Ans. Connections of all the metallic portion of the electrical equipment except the current-carrying conductors to earthing electrode are known as equipment earthing.

Q. **What do you understand by the earthing electrode?**

Ans. A metal pipe or plate or conductor electrically connected to general mass of earth is known as earthing electrode.

Q. **Where this earthing electrode is erected?**

Ans. It is erected and shall be provided at generating station, substations or consumer's premises.

Q. **In earthing system, what ECC denotes?**

Ans. It is Earth Continuity Conductor.

Q. **What is the best value of earth resistance?**

Ans. It is zero ohm.

Q. **If a system has more resistance how the earth resistance can be brought down?**

Ans. It can be brought down by any of the following methods : pouring water in earth pit, by maintaining moisture around the earth electrode, treating the earth pit by salt water, increasing the depth of pipe into ground, making more earth pits and connecting the earthing system in parallel.

Q. **Which facility the earthing provides?**

Ans. Easy flow of faulty current through the earth.

Q. **What should be the limiting value of sum of the resistances of general mass of earth and earth continuity conductor for earth fault protection?**

Ans. It should be such low that the faulty current is at least 3.5 times or more than the fuse rating, so that fuse may blow promptly.

Q. **Is there any relationship between the cross-sectional areas of the cable used in the domestic wiring and earth continuity conductors?**

Ans. The size of the earth continuity conductor should not be less than the half of the largest current-carrying conductor used in domestic installation.

Q. **If you are using aluminium strip for earthing how will you calculate the area of the earth strip?**

Ans. It should only be calculated on the basis of the equivalent size of the copper current-carrying conductor size.

Q. **If the currents carried by different installations are 15 A, 20 A, 30 A and 100 A, what size of wire will be used?**

Ans. It is SWG No. 14, 12, 10 and 8, respectively.

Q. **What is the general recommendation of the earth continuity conductor used in domestic installation?**

Ans. It is less than one ohm.

Q. **In case the length of a single conductor is more (resulting more than one ohm resistance) what should be done?**

Ans. In that case the other conductor of same length having larger cross-sectional area will be preferred because the resistance is inversely proportional to the area of cross-section.

Q. **Which instrument is used to measure earth resistance?**

Ans. Earth tester.

Q. **Protection of earth wire is essential but at a particular point it is very much important, name the point?**

Ans. Earth wire shall be protected against mechanical damages and possibility of corrosion, particularly at the point of connections to earth electrode and the earth continuity conductor.

Q. **What do you call, the pole or line of an earthed circuit which is connected to earth.**

Ans. Earthed pole.

Q. **Whom do you call earthing lead?**

Ans. The conductor by which the connection to the earth electrode is made is known as earthing lead.

Q. **In earthing process we come across two types of insulation, the functional and supplementary insulations. What do you understand by functional insulation?**

Ans. It is the actual insulation which is needed for proper functioning of equipment and is the basic protection against electric shock too.

Q. **Whom do you call supplementary insulation?**

Ans. It is called as protective insulation. It is an additional independent insulation in order to ensure protection against electric shock in case of failure of the functional insulation.

Q. **What do you call the current having a passage of electric current other than the predetermined or desired due to imperfect insulation?**

Ans. A leakage current.

Q. **What is the permissible leakage current?**

Ans. The limiting value can be stated as

$$\text{Full load current of installation} \times \frac{1}{5000} \text{ amp.}$$

PIPE EARTHING

Q. **What is pipe earthing?**

Ans. In this type of earthing a GI pipe of 1½″ diameter and having a length of 6½ ft. long is used as earth electrode. This pipe is buried under the earth, surrounded by alternate layers of salt and charcoal.

Q. **What is the thickness of layer usually preferred?**

Ans. 7.5 cm charcoal and 7.5 cm of salt alternately placed.

Q. **Where and for what purpose the funnel is used?**

Ans. It is situated in the concrete box and placed at the top of the pipe. The water is poured in the funnel to maintained the moisture around the pipe in the charcoal and salt layers.

Q. **What is the general dimension of the concrete box?**

Ans. Generally it is made 30 cm × 30 cm which is covered with cast iron cover.

Q. **Why the holes are provided in GI pipe of 1½″ dia?**

Ans. Just to release the water to surroundings salt and charcoal layers to maintain the moisture.

Q. **How the holes are provided for effective release of water?**

Ans. Diagonally and alternately.

PLATE EARTHING

Q. **What is plate earthing?**

Ans. In this system of earthing, a plate of cast iron or copper is buried and used as earth electrode, so it is known as plate earthing.

Q. **Generally what is the depth of the plate in case of plate earthing?**

Ans. 3 metres below the ground level.

Q. **What is done to maintain a good earth continuity?**

Ans. The plate is surrounded by the layers of charcoal and salt.

Q. **How the moisture is maintained?**

Ans. A pipe is arranged on one side of which it has a funnel and on other side it enters the salt and charcoal layer. Water is poured in funnel periodically to maintain the moisture.

Q. **How the earth electrode is brought?**

Ans. A conductor is lugged or riveted with the plate and brought out through a pipe.

Q. **What is the position of the plate inside the earth surface?**

Ans. It is vertically kept not horizontally.

Q. **Why the plate is buried vertically and not horizontally?**

Ans. If by chance an earthquake or other natural calamity comes and shakes the earth's homogeneous layers, the plate will be left alone if it is placed horizontally. So it is kept vertically to be in contact with so many layers and thus the plate will always be in contact to maintain the good and continuous earth connections.

Q. **What is the size of copper plate up to 30 HP, up to 100 HP, and above 100 HP?**

Ans.

Up to 30 HP	30 cm × 30 cm × 3.12 mm
Up to 100 HP	60 cm × 60 cm × 3.12 mm
Above 100 HP	90 cm × 90 cm × 3.12 mm

Q. **What are the different sizes of the cast iron plate up to 30 HP, up to 100 HP and above 100 HP?**

Ans.

Up to 30 HP	30 cm × 30 cm × 6.35 mm
Up to 100 HP	60 cm × 60 cm × 6.35 mm
Above 100 HP	90 cm × 90 cm × 6.35 mm

Q. **What is the size of earth wire up to 10 HP?**

Ans. 8 SWG copper conductor, 6 SWG copper conductor and ½″ × 1⁄10″ copper strip.

Q. **What should be the minimum possible distance between two plates, if double earthing is done?**

Ans. At least 8′ between two plates.

Q. **What is the earth resistance of ordinary soil and rocky soil?**

Ans. In no case it should exceed more then 5 Ω and for rocky soil 8 Ω.

Q. **Name the instrument used for measuring the earth resistance?**

Ans. Earth tester.

Q. **To maintain a good earthing what periodic arrangement is done?**

Ans. Regularly and periodically add water to maintain moisture around the pipe.

OBJECTIVE TYPE MULTIPLE CHOICE QUESTIONS

1. The characteristic of a good conductor is ......
   (a) to allow the free electrons to flow at low voltage,
   (b) to allow the free electrons to flow without voltage,
   (c) to offer hindrance in the flow of free electrons.
   (d) not to allow the free electrons to move even after the application of voltage.
2. ...... is the best conductor.
   (a) Gold (b) Silver
   (c) Copper (d) Aluminium
3. The impregnated paper is used as an insulator in ......
   (a) between the armature stampings
   (b) high voltage cables
   (c) commutator
   (d) HT line.
4. Mica is used for ......
   (a) preparing the heating element
   (b) preparing heating elements of room heaters
   (c) soldering
   (d) small magnets.
5. In VIR wire V, I and R stands for ......
   (a) very important rubber
   (b) vulcanised insulated resin
   (c) vulcanised Indian rubber
   (d) vulcanised impregnated role

6. In PVC cables P, V and C stands for ......
   (a) poly vinyl chloride.
   (b) poly vulcanised conductor
   (c) poly vulcanised covering
   (d) plastic vulcanised covering
7. Generally ...... wiring is used for domestic purpose.
   (a) batten (b) cleat
   (c) lead covered (d) busbar
8. ...... are used in cleat wiring.
   (a) Cleat, screws, VIR wire etc.
   (b) Cleat, batten, VIR wire etc.
   (c) Cleat, PVC wire etc.
   (d) None of the above.
9. The ...... is drawn after fixing the clips on the wooden batten.
   (a) VIR wire (b) PVC wire
   (c) flexible wire (d) PILDUSTA cable.
10. Switch should be connected on ......
    (a) neutral wire (b) phase wire
    (c) earth wire (d) half wire.
11. The batten holder is used ......
    (a) in hanging position
    (b) hung near the bed
    (c) fitted on the flat surface
    (d) to obtain the light at a particular angle.
12. In surface conduit pipe wiring the earth wire is taken ......
    (a) along the conduit pipe internally
    (b) along the conduit pipe externally
    (c) separately under the plaster
    (d) by insulating the pipe.
13. The insulation resistance test is carried out by the instrument ......
    (a) ammeter (b) voltmeter.
    (c) megger (d) energymeter.
14. In case of earth fault in any circuit the megger will indicate ......
    (a) zero (b) 50 MΩ
    (c) infinity (d) 1 MΩ.
15. If there is a break in the circuit, the megger will show ......
    (a) zero (b) infinity
    (c) 50 MΩ (d) 1 MΩ.

16. The installation can be energised if it has the insulation resistance ......
    (a) not exceeding 1000 Ω (b) not less than 1000 Ω
    (c) not exceeding 1 MΩ (d) not less than 1 MΩ
17. If the pipe is used as earth electrode the earthing is said to be a ......
    (a) pipe earthing (b) plate earthing
    (c) tap earthing (d) None of the above.
18. Plate earthing is generally used in ......
    (a) house wiring (b) industrial installation
    (c) temporary wiring (d) cleat wiring.
19. The earth resistance is measured by ......
    (a) earth tester (b) megger
    (c) voltmeter (d) ohmmeter.
20. 8 SWG wire is ...... than 16 SWG wire.
    (a) thicker (b) thinner
    (c) equal (d) None of the above.
21. Which type of wiring is done in a workshop?
    (a) Conduit pipe wiring (b) Cleat wiring
    (c) Batten wiring (d) LCC wiring
22. The ...... wiring is suitable for damp places.
    (a) batten wiring (b) cleat wiring
    (c) casing and capping wiring (d) conduit pipe wiring.
23. Which type of wiring has minimum span of life?
    (a) Conduit pipe wiring (b) Cleat wiring
    (c) Batten wiring (d) LCC wiring.
24. The advantage of using conduit pipe wiring is that it ......
    (a) is lighter (b) is stronger
    (c) has long life
    (d) resists mechanical damage and has long life.
25. The non-metallic conduit wiring is ......
    (a) easy to install (b) light in weight
    (c) corrosion resistant
    (d) easy to install, light in weight, corrosion resistant, flame retardant and impact resistant also.

CHAPTER 11

Chemical Effect of Electric Current

INTRODUCTION

Q. **What do you understand by the chemical effect of electric current?**

Ans. Whenever an electric current is passed through a solution or electrolyte, it decomposes into its ions i.e., it experiences some chemical changes. This effect is known as the chemical effect of current.

Q. **What is meant by the liquid conductor or electrolytes?**

Ans. Electrolyte is a solution or liquid conductor which undergoes a chemical change when an electric current is passed through it, for example H_2SO_4, H_2O, NaCl etc.

Q. **What are electrodes?**

Ans. The electrodes are the metallic plates which are immersed in an electrolyte. These are of two types—anode and cathode.

Q. **What is cathode and anode?**

Ans. The electrode which is connected with the negative terminal of a battery is called cathode and the electrode which is connected with positive terminal is called anode.

Q. **What is an ion?**

Ans. The charged particles of a solution are known as ions.

Q. **Define cation and anion?**

Ans. The particles which has positive charge are known as cation and the other having negative charge are known as anion.

ELECTROLYSIS

Q. **What is electrolysis?**

Ans. Electrolysis is a process in which an electrolyte decomposes into its ions.

Q. **What do you mean by the quantity of electricity?**

Ans. It is the product of current and time through which current is passed.

Q. **What are the Faraday's laws of electrolysis?**

Ans. Faraday introduced two laws of electrolysis :

1st law : The weight of the liberated ions during electrolysis is directly proportional to the quantity of electricity passed through it.

2nd law : Whenever the same quantity of electricity is passed through different electrolytes, the amount of different substance liberated is directly proportional to their chemical equivalent.

Q. **What is the mathematical representation of the first law?**

Ans.

$$m \propto q$$

where m = mass of the liberated ions

q = quantity of electricity = it

i = current, and

t = time in seconds

$$\therefore \quad m = Zit$$

where Z is a constant, known as electrochemical equivalent of the substance.

Q. **What is the second law representation?**

Ans. If there are two substances and their liberated masses are m_1 and m_2 when say I ampere current is passed for t sec then

$$m_1 \propto Z_1$$

$$m_2 \propto Z_2$$

where Z_1 and Z_2 are the electrochemical equivalents of the substances, respectively.

Q. **Where the phenomenon of electrolysis is used?**

Ans. It is used for purification of metals, electroplating and electrotyping etc.

Q. **What is electroplating?**

Ans. Electroplating is a process by which a metal is deposited on any other metal by passing the current through an electrolyte.

Q. **Is it necessary to use the electrolyte of the same material of which the electroplating is to be done?**

Ans. Yes, the metal to be deposited must be present in the electrolyte.

Q. **Which polarity does the material to be deposited has?**

Ans. Cathode.

Q. **For copper plating, which electrolyte may be taken?**

Ans. Copper sulphate solution.

Q. **If the high value of current is passed in the electrolyte, will the electroplating be smooth, fine and uniform?**

Ans. No.

Q. **What is the chemical equivalent?**

Ans. It is the mass of the element which combines or replaces one gram of hydrogen by chemical action.

Q. **What is the electrochemical equivalent?**

Ans. It is the mass of an element which is liberated by one ampere per second or one coulomb of electricity.

Q. **What do you understand by electrotyping?**

Ans. In this process, first of all the written material is prepared, and the impressions are made with the wax or plaster of paris. To make it a conductor, it is coated with powdered graphite. Then it is kept on cathode and electroplating is done, this plating will strengthen the film of impressions and the film is further strengthened by filling the back with molten metal. These electroblocks are used for printing processes.

Q. **For silver plating which electrolyte is generally used?**

Ans. Silver nitrate ($AgNO_3$).

Q. **What are ECE for Ni, Ag, Al?**

Ans. 0.000304 gm/°C, 0.001118 gm/°C, 0.000094 gm/°C, respectively.

CELLS

Q. **What are the types of cells?**

Ans. Primary and secondary.

Q. **What are primary cells?**

Ans. The cells which are able to deliver the current as soon as its components are gathered. Once discharged cannot be recharged again.

Q. **Give some names of the primary cells?**

Ans. Voltaic cell, Leclanche cell, dry cell, Bunsen cell and Bicromate cell etc.

Q. **What do you understand by secondary cells?**

Ans. The cells which store electrical energy in the form of chemical form are known as secondary cells. This cell does not deliver the current as soon as it is assembled and will deliver when it is charged. The material can be used again and again.

PRIMARY CELLS

Q. **What are the characteristics of primary cells?**

Ans. The cell once discharged cannot be recharged again, the initial cost is low, it is portable also, should have low internal resistance and should have long life.

VOLTAIC CELL

Q. **What are the different components of the voltaic cells?**

Ans. Glass container, Zn plate, Cu plate, dil. H_2SO_4, electrolyte and arrangement for checking the voltage produced.

Q. **What are the polarities of copper plate and zinc plates in this cell?**

Ans. The copper plate behaves like positive plate and zinc plate as negative plate.

Q. **What is the chemical action which takes place in this cell?**

Ans. $Zn + H_2SO_4 \rightarrow ZnSO_4 + 2H$

$2H + 2e \rightarrow H_2$

Q. **How the charge is obtained by copper plate?**

Ans. The hydrogen particles which are actually in the form of nascent hydrogen form give up their charge to copper plate and become the atomic hydrogen and thus the copper plate behaves as anode at a higher potential than zinc plate.

Q. **At N.T.P. the gases behave like a conductor or insulator?**

Ans. Insulator.

Q. **After giving the charge to copper plate will the hydrogen atom behave like a conductor or insulator?**

Ans. Insulator.

Q. **What are the defects of the voltaic cell?**

Ans. Two—the local action and polarization.

Q. **What is the local action?**

Ans. The hydrogen so formed in the chemical action should give away the charge to some opposite polarity material. The zinc used being a commercial one, has so many impurities like Cu, As, Sn, Fe, Co, Ae etc., so the hydrogen gives away the charge and forms a complete cell locally. Thus the charged particles will not go to Cu plate and cell will stop producing e.m.f. This defect is called local action.

Q. **What is polarization?**

Ans. Forming the layer of atomic hydrogen (after giving the charge) on the copper plate and not allowing the charged particles to give away the charge to copper plate, causes the reduction in e.m.f. of the cell. This defect is known as the polarization.

Q. **How the local action defect is removed?**

Ans. By amalgamating the zinc rod.

Q. **What actually is done in case of amalgamation?**

Ans. We coat a mercury layer on the zinc rod.

Q. **How does this mercury layer minimise or remove the local action?**

Ans. It only allows the zinc particles to react and the other particles are covered i.e. they do not take any part in the chemical reaction of the cell hence overcoming local action.

Q. **What are the methods of minimising the polarisation?**

Ans.
1. Manual method
2. By using the depolariser.

Q. **How it minimised by manual method?**

Ans. In this case, the zinc rod is taken out and brushed out by the manual operation, thus the hydrogen bubbles are washed out and the rod is now again ready.

Q. **How the depolariser helps in minimising the depolarization?**

Ans. It converts the hydrogen atoms into water or some other form and does not allow to assemble around copper plate.

LECLANCHE CELL

Q. **Name the electrodes used in Leclanche cell?**

Ans. Carbon and zinc.

Q. **What is the polarity of carbon plate?**

Ans. Positive.

Q. **Which plate has negative polarity?**

Ans. Zinc.

Q. **Name some important parts of the Leclanche cell?**

Ans. Glass vessel, Zn rod, carbon rod, mixture of MnO_2 and graphite, solution of NH_4Cl, porous pot etc.

Q. **What do you know about the porous pot?**

Ans. It is a pot made of the porous material. It only allows the gas particles to pass and not the liquid.

Q. **What is the chemical reaction of this cell?**

Ans. $Zn + 2NH_4Cl \rightarrow ZnCl_2 + 2NH_3 + 2H$

$2H + 2MnO_2 \rightarrow Mn_2O_3 + H_2O$.

Q. **What is MnO_2?**

Ans. It is depolariser.

Q. **How it helps in this cell for minimising the polarization?**

Ans. By converting the hydrogen particles into water.

Q. **Does this cell also have local action defect?**

Ans. Yes, because zinc rod is also used.

Q. **What is the e.m.f. of this cell?**

Ans. 1.46 volts which comes to 1 volt after some time.

Q. **Where this cell is used?**

Ans. In laboratories, and other places where continuously e.m.f. is not needed, i.e. for intermittent use like electric bell, buzzer etc.

DRY CELL

Q. **What is dry cell?**

Ans. The cell in which the active material is in the form of liquid stage, not in solution form, are known as the dry cell. These are portable and easy to use. It is the modification of Leclanche cell.

Q. **What are the constructional parts?**

Ans. Carbon rod, zinc cylinder, MnO_2 and NH_4Cl in liquefied stage and a membrane etc.

Q. **What is the polarity of zinc and carbon rod?**

Ans. Zinc cylinder behaves like a negative and carbon rod as positive terminal.

Q. **What is the chemical reaction of the dry cell?**

Ans. $Zn + 2NH_4Cl \rightarrow ZnCl_2 + 2NH_3 + 2H$

$2H + 2MnO_2 \rightarrow Mn_2O_3 + H_2O$

Q. **Is there any exit for the formed H_2O and if not what effect does it have on the performance of the cell?**

Ans. After using it for some time, the water particles collect and cell stops working and the solution may leak out breaking the zinc container.

Q. **Is there any other provision made to release hydrogen gas?**

Ans. Yes, sometimes a hole is provided on the top of the cell.

Q. **What is the e.m.f. of this cell?**

Ans. It is about 1.5 V.

Q. **What are the uses of this cell?**

Ans. It is widely used in the torch, transistors, radios, tape recorders, instruments and other electrical appliances.

BUNSEN CELL

Q. **What are the parts of the Bunsen cell?**

Ans. Carbon rod, conc. HNO_3, Zn cylinder, glass vessel, dil. H_2SO_4, and porous pot etc.

Q. **What is the depolariser in this cell?**

Ans. HNO_3.

Q. **What is the chemical reaction of this cell?**

Ans. $Zn + H_2SO_4 \rightarrow ZnSO_4 + H_2$

$H_2 + 2HNO_3 \rightarrow 2H_2O + 2NO_3$

Q. **What is the e.m.f. of this cell?**
Ans. 1.9 V and fairly constant.

Q. **Where it is used?**
Ans. In laboratories.

Q. **Why the use of this cell is limited when it has fairly constant e.m.f.?**
Ans. The gas NO_2 is a harmful gas and may cause danger to someone's life, so its use is limited and restricted.

Q. **Can you name any single cell in which two acids are used?**
Ans. Yes, it is Bunsen Cell.

BICHROMATE CELL

Q. **What are the constructional parts of the bichromate cell?**
Ans. Glass vessel, amalgamated zinc rod, mixture solution of H_2SO_4 + $K_2Cr_2O_7$, carbon plates, terminals and cork etc.

Q. **Which is the positive terminal of the cell?**
Ans. The carbon plates.

Q. **Why it is called as bichromate cell?**
Ans. Because potassium bichromate solution is used in this cell.

Q. **What is the e.m.f. of this cell?**
Ans. 1.9 to 2 V which comes down to 1.7 V.

Q. **What is the internal resistance of this cell?**
Ans. It has very low internal resistance.

Q. **Where is it used?**
Ans. Because of the rapid reduction in its e.m.f. it is not used much. It can be used in telephones and telegraphic purposes etc.

DANIELL CELL

Q. **What are the different constructional parts of the Daniell cell?**
Ans. The copper container, saturated solution of copper sulphate, porous pot, dil. H_2SO_4, and Zn rod etc.

Q. **What are the polarities of Zn rod and copper container?**
Ans. Copper container as positive and Zn rod as negative.

Q. **What is the chemical reaction?**
Ans. $Zn + H_2SO_4 \rightarrow ZnSO_4 + H_2$
$H2 + CuSO4 \rightarrow Cu + H2SO4$

Q. **Where the copper particles are collected?**
Ans. At the bottom of the container.

Q. **Which solution out of $CuSO_4$ and dil. H_2SO_4 is light?**

Ans. Dil. H_2SO_4.

Q. **Can you name the cell in which we are using the principle of gravity?**

Ans. Yes, it is Daniell cell.

Q. **How you preserve the $CuSO_4$?**

Ans. There are shelves provided near the top of the container and the $CuSO_4$ crystals are kept there.

Q. **Where this cell is used?**

Ans. In laboratories, railways signals, telephones etc.

Q. **What is the e.m.f. of this cell?**

Ans. 1.1 V.

Q. **Name the depolariser used in this cell?**

Ans. $CuSO_4$ solution.

Q. **Is there any other name of this cell?**

Ans. Yes, gravity cell.

WESTON CADMIUM CELL OR STANDARD CELL

Q. **Why this cell is called Standard cell?**

Ans. Because it gives the constant e.m.f. at a particular temperature.

Q. **What are the constructional parts of this cell?**

Ans. Wooden box, terminals, glass tube, leads or connected wires, pure distilled mercury, amalgamated cadmium, a mixture obtained by grinding the pure mercurous sulphate, a cadmium sulphate and mercury in the ratio of 8 : 4 : 1. and dissolved in the solution of cadmium sulphate, crystals of $CdSO_4$ and saturated solution of cadmium sulphate.

Q. **What is the shape of the tube used in this cell?**

Ans. H shape.

Q. **What depolariser is used in this cell?**

Ans. It is a mixture obtained by grinding the pure mercurous sulphate, cadmium sulphate and mercury in the ratio of 8 : 4 : 1 and dissolved in the solution of cadmium sulphate.

Q. **Why $CdSO_4$ crystals are placed?**

Ans. To maintain the saturation of the $CdSO_4$ solution.

Q. **Which terminal has positive polarity?**

Ans. It is pure mercury which acts as positive electrode.

Q. **Which is negative terminal of the cell?**

Ans. It is 12.5% cadmium amalgamation obtained by mixing 1 g of cadmium to 7 g of mercury filled in other tube and it works as negative terminal of the cell.

Q. **What arrangement is done to maintain the constant temperature?**

Ans. A wooden box is prepared where tube is placed and an insulation (wooden) is inserted to avoid any thermal leakage.

Q. **What is the e.m.f. of this cell.**

Ans. 1.0184 volts. at 20°C.

Q. **Is it sensitive to the temperature change?**

Ans. Yes, the e.m.f. decreases with the increase in temperature.

Q. **How much change results?**

Ans. It is 0.0004 volt/degree rise in temperature.

Q. **If the temperature decreases below 20°C will the e.m.f. increase?**

Ans. No.

Q. **What is the internal resistance of the cell?**

Ans. It is very high, 125 Ω and increases with the increase in temperature.

Q. **Can you draw more current and if no then what is the safe current limit?**

Ans. No, the safe limit is 0.0001 A.

Q. **If you draw more current than 0.0001 A, what effect will be experienced?**

Ans. The cell will discharge immediately.

Q. **What precautions should be taken for the use of this cell?**

Ans.

1. It should not be overturned.
2. More current is restricted.
3. If the e.m.f. is slightly less, keep it unused for some time to restore.
4. The operating temperature should be maintained at 20°C.
5. The material used in this cell should be of highest purity.

Q. **What is the use of this cell?**

Ans. It is used for standardization of other cells, used for research work and precise work and experiments.

SECONDARY CELL

Q. **What are the characteristics of the secondary cells?**

Ans.

1. Complete reversible reaction.
2. Internal resistance should be low.
3. Fairly constant e.m.f.
4. Losses should be minimum.
5. Good efficiency.

6. Low cost and large storage capacity.
7. Long life.

Q. **What are the types of the secondary cell?**

Ans. Lead acid cell and nickel-iron or alkaline cell.

LEAD ACID BATTERY

Q. **What is a battery?**

Ans. The combination of cells is known as battery.

Q. **What are the parts of the lead acid battery?**

Ans. Container, positive and negative plates, separators, cell cover, vent plugs, plate connector, cell connectors, sealing compound, mud space and electrolyte etc.

Q. **What is the function of the container?**

Ans. It accommodates all the parts of the cell or battery.

Q. **The container is made of which material?**

Ans. Hard rubber or bituminous compound.

Q. **How the cells are placed in battery?**

Ans. These are placed in chambers in the container?

Q. **How many chambers are there in a container?**

Ans. It depends upon the voltage of the battery or the number of cells to be connected. Generally three or six cells are connected in series. The same is the number of chambers.

Q. **According to the construction how the plates are divided?**

Ans. (i) Plante type. (ii) Faure type.

Q. **How the Plante type plates are prepared?**

Ans. These are prepared from pure lead by several charges and discharges.

Q. **What is the active material of positive plates?**

Ans. PbO_2.

Q. **What is the active material of negative plates?**

Ans. Sponge lead.

Q. **How the plates are prepared according to Faure method?**

Ans. These are the pasted plates and tubular plates. In this system a grid of lead is formed and if the active material is in the shape of paste, it is pasted on the grid and if in packets, packed in the grid and pressed, it is tubular plate.

Q. **Is it necessary to use separators?**

Ans. Yes.

Q. **Why it is necessary?**

Ans. To avoid short-circuiting between negative and positive plates.

Q. **Are both side of the separator smooth?**

Ans. No, one side is smooth, the other side is having grooves.

Q. **Name the material used for preparing the separator?**

Ans. Hard rubber, ebonite, fibre glass or wood etc.

Q. **Why the grooves are provided on one side of the separator?**

Ans. So that the electrolyte may come in contact with the active material of the plates.

Q. **Which side of the separator is kept towards the positive plate?**

Ans. Grooved side.

Q. **How many separators are used in a cell equal to positive plates or negative plates or both plates or none of these?**

Ans. Double the number of positive plates or one less than the total number of positive and negative plates.

Q. **Which electrolyte is used in lead acid cell?**

Ans. Dil. H_2SO_4.

Q. **What is the gravity of the electrolyte in case of charged and discharged condition of the cell?**

Ans. 1.280 in charged condition and 1.180 in case of discharged condition.

Q. **Name the instruments used for measuring the specific gravity of the electrolyte?**

Ans. Hydrometer.

Q. **Name the instrument used for testing the cell?**

Ans. Cell tester.

Q. **What is the chemical reaction during charging and discharging the cell?**

Ans. In discharging :

$PbO_2 + 2H_2SO_4 + Pb \rightarrow PbSO_4 + 2H_2O + PbSO_4$

In charging :

$PbSO_4 + 2H_2O + PbSO_4 \rightarrow PbO_2 + 2H_2SO_4 + Pb$

Q. **What is the e.m.f. of the cell?**

Ans. It is 2.2 volt cell.

Q. **Can this cell be recharged?**

Ans. Yes.

Q. **When the cell should be sent for recharging?**

Ans. As soon as the voltage per cell reaches to 1.8 volt or gravity of electrolyte reaches to 1.180.

Q. **What is the main factor which depends upon the gravity of electrolyte?**

Ans. Internal resistance of the cell.

Q. **What are the (i) charging p.d. (ii) fully charged p.d. (iii) fully discharged p.d.?**

Ans. (i) 2.5 to 2.75 V/cell.
(ii) 2.2 to 2.5 V/cell.
(iii) 1.7 to 1.1 V/cell.

Q. **What is topping up and gassing?**

Ans. To maintain the level of electrolyte up to ½″, above the plates is called topping up (add distilled water). When the cell is fully charged gases come out from the cell, it is known as gassing.

Q. **What are the different methods of charging the L.A. battery or cell?**

Ans.
1. Constant current method.
2. Constant voltage method.

Q. **What are the types of charging according to the charging rate?**

Ans. Initial charging, normal charging, boosting charging, trickle charging and equalizing charging.

Q. **What is boosting charging?**

Ans. The charging which is done up to 6 times the normal charging rate. It is done when the battery is required within three to four hours only.

Q. **What is the type of charging which is done to remove the sulphation of the plates?**

Ans. Equalizing charging.

Q. **What is the current rate generally kept in case of trickle charging?**

Ans. 1/50th of normal charging current.

Q. **Which type of rectifier is used in case of battery charger?**

Ans. Either metallic rectifier or tunger bulb rectifier.

Q. **What are the types of efficiencies of L.A. cell?**

Ans. Ampere-hour and watt-hour efficiency.

Q. **What is ampere-hour efficiency?**

Ans. It is the ratio of total ampere-hour supplied during discharging to the total ampere-hours given during charging, generally expressed in percentage.

Q. **What are the defects in L.A. cell?**

Ans. Sulphation, buckling, sedimentation, internal short circuit and loss of electrolyte.

Q. **What is buckling, and loss of electrolyte?**

Ans. Deformation of plates is known as buckling. For its remedy take out the plate, make them straight by mechanical means and if it is in usable condition use it, otherwise replace the plates.

Q. **Why the leads to a battery while it is on for charging be disconnected?**

Ans. Because the spark may occur and may ignite the gas and cause an explosion.

Q. **Should we keep the same frequency of testing an accumulator in different weather?**

Ans. No, test the storage battery more frequently in cold weather than in summer. Often the discharged battery freezes easily.

Q. **When a battery is said to be floating?**

Ans. The condition in which it supplies current intermittently and also during off cycle it gets charged.

Q. **Generally the gravity of electrolyte of L.A. cell is given as 1.280, but how do we call it in the market?**

Ans. In the market it is termed as 1280 in charged condition and 1180 in discharged condition.

NICKEL-IRON ALKALINE CELL

Q. **What is the other name generally given to this cell?**

Ans. It is known as Ni-Fe cell, Edition cell and Jungler cell.

Q. **Which is the active material of the positive plate?**

Ans. NiO_2.

Q. **Name the active material of negative plate?**

Ans. Fe.

Q. **What are the main parts of the Ni-Fe cell?**

Ans. Container plates, separators, electrolyte.

Q. **The container is made of which material?**

Ans. Container is made of nickel-plated steel box.

Q. **The separators are made of which material?**

Ans. Ebonite.

Q. **Name the electrolyte used in this cell?**

Ans. Mixture of $KOH + H_2O$ in the ratio of 21 : 79, respectively.

Q. **Why should be the level of electrolyte above the plates?**

Ans. From ½″ to 3″ above the plates.

Q. **What is the chemical equation of the cell?**

Ans. $$\underset{+\text{ plate}}{NiO_2} + 8KOH + H_2O + 3Fe \rightarrow \underset{-\text{ plate}}{2NiO_4} + 8KOH + H_2O + Fe_3O_4$$

Q. **What is the specific gravity of the electrolyte?**

Ans. 1.230.

Q. **What is the specific gravity during charging and discharging of the cell?**

Ans. It remains unchanged.

Q. **What is the e.m.f. of the cell under :**

(i) fully charged condition.

(ii) fully discharged condition.

(iii) recharging the cell.

Ans. (i) 1.4 V/cell. (ii) 1 V/cell. (iii) 1.7 V/cell.

Q. **What approximation of the internal resistance of LA cell and Ni-Fe cell is?**

Ans. It is five times more in Ni-Fe cell than the L.A. cell.

Q. **How much durable this cell is?**

Ans. It is very much durable and can be used in any condition, even can be short circuited.

Q. **What are the charging methods of the cell? Are these different from the L.A. cell?**

Ans. No, the same as that of L.A. cell.

Q. **Where this cell is used?**

Ans. It is used in aeroplanes, ships, locomotives, and industrial trolleys etc.

Q. **Which cell has got more capacity for the same frame size Ni-Fe cell or L.A. cell?**

Ans. Ni-Fe cell.

OBJECTIVE TYPE MULTIPLE CHOICE QUESTIONS

1. The voltage of primary cell depends upon ......

   (a) area of the plate (b) distance between the plates

   (c) electrolyte and types of plates (d) thickness of the plate.

2. In case of primary cell ......

   (a) chemical energy is converted into mechanical energy.

   (b) chemical energy into electrical energy.

   (c) electrical energy into chemical energy.

   (d) electrical energy into mechanical energy.

3. ...... are the examples of primary cells.

   (a) Voltaic cell, Leclanche cell, dry cell

(b) Voltaic cell, Leclanche cell, alkaline cell
(c) Dry cell, Leclanche cell, alkaline cell
(d) Lead acid, alkaline cell, standard cell

4. The local action is minimized by amalgamating the ......
(a) zinc rod (b) carbon or copper rod
(c) container (d) terminals

5. There are two defects of the voltaic cell ......
(a) sulphation and buckling (b) buckling and sedimentation
(c) polarisation and local action (d) local action and buckling.

6. The positive electrode of dry cell is made of ......
(a) copper (b) carbon (C) zinc (d) sulphur

7. ...... has 1.9 volt e.m.f.
(a) voltaic cell (b) dry cell
(c) Leclanche (d) Bunsen cell.

8. Generally the Leclanche cell is used for ...... purposes.
(a) continuous (b) heavy (c) low (d) intermittent

9. The positive terminal of bichromate cell is made of ......
(a) zinc (b) copper
(c) carbon (d) potassium.

10. Lead acid battery is generally used for ......
(a) electric bell in house (b) automobiles
(c) aeroplanes (d) trains.

11. State true or false :
(a) The positive plate of lead acid battery is of grey colour.
(b) The electrolyte used in L.A. cell is dil. sulphuric acid.
(c) Secondary cell can be recharged.
(d) The randum material of primary cells can be used again and again.
(e) Standard cell is used to take supply for electric bell.

12. Choose correct answer :
(a) The charging PD should not be more than 1.8 V/2.75 V per cell for lead acid cell.
(b) The discharged P.D. should not be allowed below 1.8/2.2 V per cell for lead acid cell.
(c) The specific gravity of the electrolytes should be 1180/1280.
(d) The mud space/cell connector is used for accumulating the falling active material.
(e) The separator/electrolyte is used to avoid short circuit between the negative and positive plates.

(f) The deformation of plates is known as sedimentation/buckling.

(g) The charging of battery in a short interval is known as initial/boosting charging.

(h) The vent plugs allow to bypass electrolyte/gases.

(i) For battery charging we require A.C./D.C. supply.

(j) While charging, the specific gravity of electrolyte increases/decreases.

13. Correct the sentences :

(a) In charging process the voltage per cell increases/decreases.

(b) In charging the chemical energy/electrical energy is stored in electrical/chemical energy.

(c) Local action/sedimentation is the defect of simple voltaic cell.

(d) Zn rod is amalgamated/insulated in case of Leclanche cell.

(e) Bunsen/Daniell is also known as the double fluid cell.

14. Which of the following statement is false :

(a) Lead acid cell can be recharged.

(b) The primary cell can be recharged.

(c) A secondary cell has completely reversible chemical reaction.

(d) The output of L.A. cell is 2 volts per cell.

(e) The flow of current in the chemical combination of the cell is from positive to negative.

(f) The positive electrode of dry cells is carbon.

(g) The container of Daniell cell is made of copper.

(h) The local action is the defect of L.A. cell.

(i) Alkaline cell has more life.

(j) $CdSO_4$ is used in standard cell.

15. Correct the conditions when a lead acid cell is to be recharged.

(a) Connect negative terminal of the battery to the positive terminal of the charging source.

(b) The battery voltage must be more than the charging unit voltage.

(c) The charging current must flow in the same direction in which the discharging current.

(d) Open the vent plugs while putting the battery on charge.

(e) Maintain the level of electrolyte about 20 mm above the plates.

16. When the secondary cells have discharged, they are generally ……

(a) recharged (b) kept idle

(c) discarded (d) put on a time tester.

17. Which cell has more life?
    (a) Voltaic cell
    (b) Ni-Fe cell
    (c) Lead acid cell
    (d) Dry cell.
18. State true or false :
    (i) The capacity of a storage cell is expressed in volts.
    (ii) The condition of fully charged battery is checked by means of hydrometer.
    (iii) Dry cell has completely reversible chemical reaction.
    (iv) The voltage generated by a silver-cadmium secondary cell is 1.1 V.
    (v) During charging the electrolyte of the lead acid battery becomes stronger.
    (vi) The term trickle charging is associated with dry cell.
    (vii) In nickel-iron cell the electrolyte is a solution of dilute potassium hydroxide.
    (viii) Electrolyte for silver plating may be silver nitrate solution.
    (ix) During float charging the voltage remains constant.
    (x) The positive terminal of a dry cell is made of carbon.

CHAPTER 12

Magnetism and Electromagnetism

INTRODUCTION

Q. **What is the magnet?**

Ans. The substance which has the property of attracting the iron, iron fillings or magnetic substances is known as magnet.

Q. **If a magnet is freely suspended, in which direction will it indicate?**

Ans. North and South.

Q. **What are the types of magnet?**

Ans. Natural and artificial.

Q. **What are the natural magnets?**

Ans. The magnets which are found in nature and have the property of attraction and direction both, are called as natural magnets.

Q. **What do you mean by the artificial magnet?**

Ans. The magnets which are made by human beings by artificial means are called artificial magnets.

Q. **What are the types of artificial magnets?**

Ans. Temporary and permanent.

Q. **Which magnets are called temporary magnets?**

Ans. The magnet which retains its magnetism as long as the magnetising force is acting, is known as the temporary magnet.

Q. **What do you understand by the permanent magnet?**

Ans. The magnets which retain their magnetism even after removing the magnetising force, permanently or for a very long time are known as permanent magnets.

Q. **Which material is used for making permanent magnets?**

Ans. Tungsten steel or high carbon steel.

Q. **Name the different shapes of permanent magnet?**

Ans. Bar magnet, horseshoe magnet, U-shape magnet and magnetic needle.

TERMS AND DEFINITIONS

Q. **What are the lines of force?**

Ans. These are imaginary curves through which a magnet is supposed to act.

OR

The path along which a unit north pole would move, if it is free to move, is known as magnetic line of force.

Q. **What are the properties of lines of force?**

Ans.
1. The line of force starts from north pole and terminates at south pole.
2. There is no perfect insulator for magnetic lines of force.
3. The lines of force never cross each other.
4. These also try to follow the shortest path.

Q. **What is magnetic attraction?**

Ans. If two opposite poles are placed near each other they will experience a force which will try to bring both poles further nearer, this phenomenon is called magnetic attraction.

Q. **Define magnetic field?**

Ans. The space occupied by the lines of force around a magnet is called magnetic field.

Q. **What is magnetic flux?**

Ans. The total number of lines of force around the magnet is called magnetic flux. It is represented by ϕ.

Q. **What is the MKS unit of magnetic flux?**

Ans. Weber.

Q. **What are the magnetic poles?**

Ans. The two ends of a magnet where the magnetic field is very much concentrated or maximum, are known as magnetic poles.

Q. **How many poles are there in a magnet?**

Ans. North and south—only two.

Q. **What is magnetic axis?**

Ans. The imaginary line joining both magnetic poles (north and south) is called magnetic axis.

Q. **Define magnetic neutral axis?**

Ans. It is also an imaginary line which right bisect the magnetic axis.

Q. **Is there any magnetism at the center point (intersecting point of both magnetic and neutral axis)?**

Ans. No.

Q. **What is magnetic repulsion?**

Ans. If two poles of same polarity are kept near then they experience a force to force them further away, that is called magnetic repulsion.

Q. **Is the force of attraction or repulsion depends upon the pole strengths?**

Ans. Yes.

Q. **State Coulomb's second law?**

Ans. The force of attraction or repulsion between two poles is directly proportional to the product of their pole strengths and inversely proportional to the square of distance between them.

Q. **State Coulomb's first law?**

Ans. Like poles repel each other and unlike poles attract each other.

Q. **What is a unit north or south pole?**

Ans. If two similar poles are placed at a distance of 1 cm apart and experiences a force of attraction or repulsion of one dyne, then the poles are said to have a unit pole strength.

Q. **Can you have an isolated pole?**

Ans. No, it cannot exist.

Q. **If a permanent magnet is broken into a number of pieces, what will happen?**

Ans. Every piece will act as an independent magnet.

Q. **What will be the force in MKS system when defining the unit pole?**

Ans. $\frac{1}{4\pi\mu_0}$ newtons in vacuum.

Q. **What is magnetic induction?**

Ans. Forming of an opposite pole on a piece of iron or steel by bringing near a magnet is known as magnetic induction.

Q. **What is permeability?**

Ans. It is the ratio of the magnetic flux produced in magnetic substance to the magnetic flux in air, by the same coil. It is denoted by μ.

Q. **What is the flux density?**

Ans. The flux per unit area is called the flux density. It is denoted by B and unit is Wb/m^2.

Q. **What are the care and maintenance of the magnet?**

Ans.
1. It should not be thrown, it should be handled properly.
2. It should not be heated up.
3. It should not be hammered.

Q. **What are the methods for making artificial magnet?**

Ans.
1. Touch method.
2. Electrical method.

Q. **How many are the ways to make artificial magnets by touch method?**

Ans. Single touch, double touch and divide and touch method.

Q. **What are the disadvantages of touch method?**

Ans.
1. Require more time.
2. The magnet will not be more strong than the magnet used.

Q. **How an electromagnet is prepared?**

Ans. A number of turns are wound over an iron piece or specimen and energised by D.C. mains of low voltage, then the specimen will behave like a magnet which will be known as electromagnet.

Q. **How the magnetic lines of force are traced?**

Ans. By means of magnetic compass of magnetic needle or iron fillings.

Q. **What are the uses of electromagnet and also tell two advantages?**

Ans. The electromagnet is used in all machines like D.C. motor, D.C. generator, A.C. motor, alternator, transformer, electric bell etc. The advantages can be said as the electromagnet can be energised in no time and the magnetic strength can be increased or decreases as desired.

Q. **Give a brief idea of the Weber's theory or molecular theory of a magnet?**

Ans. According to this theory every molecule is a magnet, if the molecules are arranged in a systematic way, the piece will behave like a magnet and if in haphazard manner, it will not behave like a magnet.

Q. **What do you mean by the saturation point?**

Ans. If all or approximately all molecules are arranged systematically and even after increasing the magnetising force, the magnetism is not increased much (negligible) the point is said to be a saturation point.

Q. **What is the magnetising force?**

Ans. The force which is used to magnetise the spacemen is known as magnetising force.

ELECTROMAGNETISM

Q. **If the electric current is passed though a conductor, will there be any magnetic field produced around the conductor?**

Ans. Yes.

Q. **What this effect of electric current is known as?**

Ans. Magnetic effect of electric current.

Q. **What is the angular displacement between the current passing through the conductor and the magnetic field produced?**

Ans. At right angles to each other.

Q. **How the direction of lines of force around the conductor is determined?**

Ans. It can be determined by the rules namely the right hand rule, screw rule, arrow head rule or end rule and the magnetic compass.

Q. **How the direction of line of force is determined by right hand rule?**

Ans. Hold the conductor in your right hand in such a way so that the thumb indicates the direction of current, then the wrapped fingers will indicate the direction of lines of force around the conductor.

Q. **What is Cork's screw rule?**

Ans. Imagine a right handed screw to be placed along the current carrying conductor then turn the screw in such a way that the tip of screw moves in the direction of current, the direction of rotation of the screw or head will indicate the direction of lines of force around the conductor.

Q. **If a wire is bent in such a way (a) only one turn (b) having more member of turns of less axial length and more diameter (c) having more turns, less diameter and more axial length, what are these known as?**

Ans. (a) Loop; (b) Coil; (c) Solenoid.

Q. **Can you find out the polarity of a solenoid if carrying current?**

Ans. Yes, by means of right hand rule.

Q. **Name the different rules for finding out the polarity of the solenoid?**

Ans. Right hand rule, end or arrow rule, screw rule and magnetic needle.

Q. **How the right hand rule is used for finding out the polarity of a solenoid?**

Ans. Grip the solenoid in your right hand in such a way so that the fingers indicating the direction of current in the solenoid, then the thumb will indicate the direction of north pole.

Q. **How end rule is applied to find out the polarity of a solenoid?**

Ans. In the solenoid, the direction of current is observed from one side. If the direction of current in the coil is clockwise the end will be forming south pole and if the direction of current in the coil is anticlockwise then the end will be forming north pole.

Q. **Can you tell what is the flux produced by the solenoid carrying *I* amperes current, having *N* turns and *L* the length?**

Ans. $\phi = B \cdot A$

$$= \frac{\mu_0 \cdot \mu_r \cdot N \cdot I}{l} \times A \text{ Wb}$$

Q. **Whenever a conductor is carrying current, the field intensity produced will be maximum at the central core of the conductor or the surface of the conductor?**

Ans. Maximum at the surface of the conductor and decreases with the increase of the distance from the surface of the conductor.

Q. **If the direction of current in the two conductors is same which force will they experience?**

Ans. Attractive.

Q. **If the direction of current in two conductors is in opposite direction which type of force is resulted?**

Ans. Repulsive.

Q. **What is meant by ampere-turn and magnetomotive force?**

Ans. The ampere-turn is the product of current flowing through the coil and the number of turns in that coil. Its symbol is AT. The m.m.f. is defined as the force which compels the magnetic lines of force to complete their circuit.

Q. **What is the relationship between flux, m.m.f. and reluctance?**

Ans. $\text{Flux} = \dfrac{\text{m.m.f.}}{\text{Reluctance}}$

Q. **What is reluctance?**

Ans. It is the opposition offered to the magnetic flux to complete its path. Its unit is AT/Wb.

Q. **What are the factors on which the reluctance depends?**

Ans. The length, area of cross-section and permeability.

Q. **What does the relationship between the reluctance, the length, area of cross-section and the permeability exist?**

Ans. $\text{Rel} = \dfrac{\text{Length}}{\text{Permeability} \times \text{Area of cross–section}}$ AT/Wb.

Q. **What is retaintivity?**

Ans. The property of retaining the magnetism by a magnetic substance is called retaintivity. The retaintivity of iron is less than the steel.

Q. **What is the residual magnetism?**

Ans. It is the magnetism left in a magnetic substance after removing the magnetising force or say m.m.f.

Q. **What is the magnetizing force?**

Ans. It is the magnetomotive force per unit length.

Q. **What are the types of substances according to the magnetic properties?**

Ans. Diamagnetic, paramagnetic and ferromagnetic.

Q. **What are the diamagnetic substances?**

Ans. The substances whose magnetic permeability is less than one are called diamagnetic substance.

Q. **What is the permeability of paramagnetic substance?**

Ans. Greater than one.

Q. **What do you know about the ferromagnetic substances?**

Ans. The material whose permeability is more than one but in very high quantity say 1000, 2000, 2500 etc., are called the ferromagnetic substances.

Q. **What is leakage factor?**

Ans. It is the ratio of total flux produced to the useful flux. It is always more than one.

MAGNETIC HYSTERESIS

Q. **What is the B.H. curve?**

Ans. The curve which shows the relationship between the magnetising force and flux density is known as B.H. curve.

Q. **What is the utility of B.H. curve?**

Ans. It indicates the permeability of the material $\mu = B/H$ and also the saturation point and knee point.

Q. **What is hysteresis?**

Ans. Lagging of magnetism behind the magnetising force is known as hysteresis.

Q. **What is hysteresis loop?**

Ans. It is the curve obtained by increasing and decreasing the positive and negative magnetizing force and the resulting flux density. It forms a closed loop, so it is known as hysteresis loop.

Q. **What do you mean by the corrosive force?**

Ans. It is the force required to remove the residual magnetism.

Q. **What is the residual magnetism?**

Ans. It is the magnetism which is left in the specimen even after removing the magnetising force.

Q. **What is the hysteresis loss?**

Ans. It is the power or the force which is used to remove the residual magnetism in the cycle of magnetisation. The corrosive force does not perform any useful work and is merely a wastage and this power wasted is known as hysteresis loss.

Q. **On what factors does the hysteresis loss depend?**

Ans. Maximum flux density, frequency or cycle of magnetisation, volume of the substance and type of material.

Q. **What does the area of loop indicate?**

Ans. The loss per cycle of magnetisation. The area of loop is wattage lost.

Q. **Which material will you recommend for A.C. machine, having large area of loop or less area of loop?**

Ans. Less area of loop, because the wattage lost per cycle of magnetisation is less.

Q. **What are the effects of hysteresis?**

Ans. More hysteresis loss, more heat developed, more energy loss and the insulation may be damaged.

Q. **How will you minimise the hysteresis loss?**

Ans. By adding silicone percentage in steel for the laminations.

ELECTROMAGNETIC INDUCTION

Q. **State the Faraday's laws of electromagnetic induction.**

Ans.
1. Whenever the flux associated or linked with an electric circuit is changed, an e.m.f. is induced in that circuit.
2. The induced e.m.f. is directly proportional to the rate of change of flux linkage.

Mathematically, $e = \frac{d\phi}{dt} \times N$ volts.

Q. **Does the e.m.f. induced depend upon the flux linkage?**

Ans. Yes, it is directly proportional.

Q. **What is Lenz's law?**

Ans. The Lenz's law states that the e.m.f. induced because of Faraday's laws of electromagnetic induction has the direction in such a way as to oppose the cause which is responsible for the production of the e.m.f. Here it will oppose the change in flux which induces it and mathematically we can say that,

$$e = -\frac{d\phi}{dt} \times N \text{ volts.}$$

Q. **On what factors does the e.m.f. induced depend?**

Ans. The rate of change of flux and the number of conductors.

Q. **If the flux density, length of the conductor and velocity is also given what is the e.m.f. induced?**

Ans. $e = B \cdot l \cdot V \cdot \sin\theta$ volts.

Q. **How the direction of induced current is found?**

Ans. By Fleming's right hand rule.

Q. **State the Fleming's right hand rule?**

Ans. According to the Fleming's right hand rule, stretch your right hand in such a way so that, the thumb, forefinger and middle finger are at right angle to each other, then if the first finger indicates the direction of field or magnetic flux, the thumb indicates the direction of motion, then the middle finger will indicate the direction of induced current.

Q. **Where this rule is applicable?**

Ans. In generators.

Q. **Where Fleming's left hand rule is applied?**

Ans. In motors to find out the direction of rotation or motion produced.

Q. **What are the types of induction?**

Ans. Static and dynamic.

Q. **What is the dynamically induced e.m.f.?**

Ans. Whenever a conductor cuts the lines of force at an angle, an e.m.f. is induced in that conductor. Here in dynamically induced e.m.f., the relative motion between the flux and the conductor is essential. It can be either rotating the magnetic field and conductor as stationary or conductor rotating and magnetic field as stationary.

Q. **How the statically induced e.m.f. is produced?**

Ans. Because of (i) self induction and (ii) mutual induction.

Q. **What is self induction?**

Ans. Whenever in a coil the current is changed or the changing flux links with the same coil the e.m.f. is induced in the same coil. This phenomenon is known as self-induction.

Q. **What is mutual induction?**

Ans. In this case the e.m.f. is induced in the coil placed neighbouring the coil, to be energised, this phenomenon is known as the mutual induction.

Q. **Give one example each of self-induction and mutual induction?**

Ans. The autotransformer is an example of self-induction and the transformers having two separate windings are the example of mutual induction.

Q. **How the flux in the transformer completes its circuit?**

Ans. Through iron cores.

Q. **If the iron cores of the transformer are replaced by wooden cores what will happen?**

Ans. Wooden articles cannot help the magnetic flux to complete its path and as a result the coil will burnt out.

Q. **What are eddy currents?**

Ans. The current produced in the undesired portion, say whenever the armature rotates in the magnetic field, the conductors and the cores both link the flux, so the e.m.f. is also induced in undesired portion and causes the currents. These are named as eddy currents.

OR

The e.m.f. produced in masses of metals whenever these metals are moved in the magnetic fields and these e.m.f. cause the short circuiting currents called the eddy currents.

Q. **Can you avoid these losses?**

Ans. No, but can be minimised.

Q. **How these losses are minimised?**

Ans. By laminating the armature.

Q. **Can you give any example wherever these eddy currents are useful?**

Ans. Yes, for damping purpose in instruments say energy meter and induction meters and in electric furnaces.

OBJECTIVE TYPE MULTIPLE CHOICE QUESTIONS

1. Magnetically all substances are divided into ......
   (a) solid, liquid and gaseous
   (b) conductor, insulator and semiconductor
   (c) ferromagnetic, paramagnetic and diamagnetic
   (d) conductor, ferromagnetic and solid.
2. Ferromagnetic substances are ......
   (a) good insulator (b) good conductor
   (c) the same as that of diamagnetic material
   (d) strongly attracted by a magnet.
3. Paramagnetic substances are ......
   (a) weakly attracted by a magnet.
   (b) the same as that of diamagnetic material.
   (c) weakly repelled by a magnet.
   (d) produced by heating iron above the curic point.
4. Diamagnetic substances are ......
   (a) magnetic monopole (b) repelled by a magnet
   (c) very good insulation
   (d) alloy of ferromagnetic substance and lead.
5. Permeability is the measure of ......
   (a) conductivity of a substance for magnetic lines of force.
   (b) the resistance of a metal
   (c) the sum of the voltage drops around a magnet
   (d) the number of electrons in a valence bond.
6. The relative permeability is the ratio of ......
   (a) resistance and current
   (b) magnetic flux to area of cross-section
   (c) magnetic flux in any medium to the magnetic flux produced in the air by the same coil
   (d) magnetic flux in any medium to the flux produced by the other coil in the vacuum.

7. The force of attraction or repulsion between two magnetic poles is ......
   (a) equal to earth's gravitational pull
   (b) directly proportional to the distance between them
   (c) equal to the magnetic field intensity multiplied by permeability.
   (d) directly proportional to their pole strength and inversely proportional to the square of distance between them.
8. A magnetic line of force ......
   (a) passes through a magnet from its north to south pole
   (b) leaves the south pole of a magnet and re-enter at its north pole
   (c) passes through a magnet from its south pole to its north pole
   (d) crosses each other
9. Weber's molecular theory states that ......
   (a) an iron bar becomes longer when magnetised
   (b) iron bar becomes shorter when magnetised
   (c) each atom is a magnet having north and south poles
   (d) current is directly proportional to voltage.
10. The atoms in a magnet are ......
   (a) arranged systematically having all the north poles of every atom in the same direction
   (b) arranged haphazardly
   (c) arranged systematically but facing south to south of all atoms
   (d) arranged systematically facing north to north of all atoms.
11. The magnetic lines of force are induced around the current-carrying conductor, this effect is known as ......
   (a) heating effect of current (b) magnetic effect of current
   (c) chemical effect of current (d) X-ray effect of current.
12. The direction of the magnetic lines of force around a current-carrying conductor can be determined by ......
   (a) right hand thumb rule (b) Fleming's right hand rule
   (c) Lenz's law (d) Boyle's law
13. The magnetic needle is ......
   (a) electromagnet (b) permanent magnet
   (c) natural magnet (d) temporary magnet
14. A solenoid is defined as an electromagnet ......
   (a) having only one turn
   (b) having more axial length than diameter
   (c) having less axial length than diameter
   (d) having more resistance.

15. The magnetic field intensity inside a magnet depends upon ......
    (a) the diameter of the coil
    (b) the dielectric coefficient of the coil
    (c) the length of the coil, number of turns and value of current
    (d) the diameter of the coil, dielectric coefficient of the cores, total number of turns and the voltage impressed.
16. To find the number of lines per unit cross-sectional area ......
    (a) multiply the number of ampere-turns per unit length by the permeability
    (b) multiply the number of volt per turn by the diameter of the coil divided by the permeability
    (c) multiply the number of ampere-turns to the coil per unit length of the conductivity
    (d) multiply the number of ampere-turns per unit length by the reluctivity.
17. The flux density is the flux divided by ......
    (a) area of cross-section
    (b) permeability
    (c) conductivity
    (d) area of the current-carrying conductor.
18. The magnetic circuits are analogous to ......
    (a) flow of permeability in an iron core
    (b) travel around a hysteresis loop
    (c) electric circuit
    (d) reluctivity which opposes the flow magnetic dipoles.
19. The m.m.f. can be compared with ......
    (a) the force of attraction between two magnetic poles
    (b) the force of repulsion between two magnetic poles
    (c) the force of earth's magnetic field
    (d) electromotive force.
20. The magnetic flux can be compared with ......
    (a) electrostatic flux
    (b) electric current
    (c) magnetic current
    (d) magnetomotive force.
21. The reluctance can be compared with ......
    (a) conductance
    (b) inductance
    (c) resistance
    (d) capacitance.
22. The magnetic circuit cannot be ......
    (a) connected in series with another magnetic circuit
    (b) connected in parallel with another magnetic circuit
    (c) continuous

(d) completely opened.

23. The reluctivity characteristic of iron is ......
    (a) the same as the resistivity
    (b) non-linear
    (c) linear
    (d) the same as that of conductivity.
24. Magnetic flux developed by an electromagnet is proportional to ......
    (a) its flux density
    (b) the square of the cross-sectional area
    (c) the reciprocal of the cross-sectional area of each pole
    (d) the square of the flux density in each pole.
25. If two reluctances are connected in series the total reluctance is equal to ......
    (a) the product of both reluctances divided by their sum
    (b) the quotient of the two reluctances
    (c) the product of these two reluctances
    (d) the sum of the two reluctances.
26. When the iron core has large air gap the total reluctance is practically the same as the reluctance of ......
    (a) the air gap divided by the reluctance of the iron core
    (b) the iron divided by the reluctance of the air gap
    (c) the iron
    (d) the air gap.
27. The material which has least area of the BH loop will have ...... hysteresis loss.
    (a) least (b) highest
28. Nickel is a ferromagnetic material.
    (a) True (b) False.
29. A low hysteresis coefficient material will have hysteresis losses of low order.
    (a) True (b) False.
30. State true or false.
    (a) Permanent magnets are made from paramagnetic substances.
    (b) Permeability of iron is less than aluminium.
    (c) Reluctance can be calculated by dividing flux by the m.m.f.
    (d) Ferrite material is used for magnetic memory purposes.
    (e) Platinum is a paramagnetic material.

CHAPTER 13

D.C. Generator

INTRODUCTION

Q. **What is a generator?**

Ans. The machine which converts the mechanical energy into electrical energy is known as the generator.

Q. **What do you understand from D.C. generator and alternator?**

Ans. The machine which converts mechanical energy into D.C. electrical energy is known as the D.C. generator and the machine which converts the mechanical energy into A.C. electrical energy is known as alternator or A.C. generator.

CONSTRUCTION

Q. **How do you classify the parts of a D.C. generator?**

Ans. Stationary parts and rotary parts.

Q. **Name the different stationary parts of the D.C. generator?**

Ans. Body or yoke, poles, side covers, bearings, eyebolt, bed sheet, rocker, terminal box, etc.

Q. **Name the different rotary parts of D.C. generator?**

Ans. Armature, shaft, pulley, commutator, fan, etc.

Q. **What is the commutator?**

Ans. The commutator is a rotating part of the D.C. machine. It is mounted on the shaft and it is having a number of copper segments which are insulated from each other and shaft.

Q. **What is the main function of the commutator?**

Ans. The main function of the commutator is to change the direction of induced A.C. e.m.f. into D.C. for external circuit.

Q. **How you collect D.C. from the commutator?**

Ans. With the help of brushes.

Q. **How commutator changes A.C. into D.C.?**

Ans. The brushes are fixed in rocker and poles are also mounted stationary so whenever the conductor comes in contact with the brushes, it has definite magnitude and direction of e.m.f. and all the conductors at this particular position will have only this e.m.f. which will be the same direction or magnitude for external circuit. Thus originally produced A.C. in the conductor is changed to D.C. for the external circuit.

Q. **Why brushes are used?**

Ans. To collect the electrical energy for external load.

Q. **Which type of materials are used for the brushes?**

Ans. Carbon, carbon + copper and sometimes purely copper.

Q. **Where carbon brushes are used?**

Ans. For low capacity machine.

Q. **What are the drawbacks of the copper brushes?**

Ans. The copper brushes are hard and may spoil the commutator segments.

Q. **How the brushes are held on the commutator?**

Ans. In rocker.

Q. **Can you change the position of the brushes?**

Ans. Yes, by rotating the rocker.

Q. **Where the maximum e.m.f. can be collected?**

Ans. At MNP.

Q. **What is MNA?**

Ans. It is the magnetic neutral axis.

Q. **What is MNP?**

Ans. Magnetic neutral plane.

Q. **Name the material generally used for yoke?**

Ans. Cast steel.

Q. **Is there any provision of lifting the machine, provided with the body?**

Ans. Yes, depending upon the weight, one or two eyebolts are provided. generally on the top of the machine.

Q. **Can the poles be solid of homogenous material?**

Ans. For very small capacity machine it can be of solid or homogeneous nature but for others it is either made of laminations of silicone steel or only the pole shoes are laminated.

Q. **How the laminations are assembled?**

Ans. The laminations of silicone steel are hydraulically compressed and then riveted to form a solid block.

Q. **Why these are made of laminations of silicone steel?**

Ans. To minimise the eddy current losses and hysteresis losses.

Q. **What is hysteresis loss?**

Ans. It is the loss due to reversal of magnetic flux in the cores.

Q. **Can you tell the formula for calculating the hysteresis loss?**

Ans. $W_h = \eta \cdot B_m^{1.6} \cdot f \cdot V$ joules/sec or watts

where W_h = hysteresis loss, f = frequency of reversal, V = volume, η = Steinmetz hysteresis constant, B_m = maximum flux density.

Q. **What is the eddy current loss?**

Ans. The losses are due to eddy currents. These losses merely produce heat and are the wastage.

Q. **Can you tell how these losses are calculated?**

Ans.

$$W_e = k \cdot B_m^2 \cdot f^2 \cdot t^2 \cdot V \text{ watts}$$

where W_e = wattage losses due to eddy currents, B_m = maximum flux density, f = frequency of magnetic reversal, t = thickness of each lamination, V = volume, k = a constant.

Q. **Is the armature made of solid or of laminations of silicone steel?**

Ans. Made of lamination of silicone steel.

Q. **How the armature is assembled?**

Ans. Assembled with the highly compressed laminations of silicone steel which are riveted to form a solid block.

Q. **What is the main advantage of making it with the laminations of silicone steel?**

Ans. To minimise the eddy current and hysteresis losses.

Q. **Where the winding is placed?**

Ans. In the slots.

Q. **What are slots and which types are generally used?**

Ans. The slots are the grooves made for housing the conductors i.e. winding. These are of three types—the open, semi-closed and totally closed type. Generally the open type slots are used for medium size machine.

Q. **Why horizontal ducts are provided?**

Ans. The horizontal ventilating ducts are provided for better cooling.

Q. **Why the covers are provided/used?**

Ans. To keep the rotor in prefect centre, to provide a covering and mechanical protection from outside.

Q. **Can you name the covers?**

Ans. Yes, front end cover and rear end cover.

Q. **Is there any additional arrangement provided for effective cooling?**

Ans. Yes, a fan is provided on the shaft.

Q. **Does the yoke carry the same flux as is the flux per pole?**

Ans. No, yoke carriers only half the flux.

Q. **Is there any necessity of the terminal box?**

Ans. Yes, for better and tight connections.

Q. **How the poles are fixed with the body?**

Ans. Screwed or bolted.

Q. **Why the pole faces are curved?**

Ans. To provide a uniform air gap around the armature cores.

Q. **What will happen, if the air gap is increased?**

Ans. The reluctance will increase, as a result if it is motor, it will take more current to establish the same torque and if generator the output will decrease.

Q. **What is the pig tail?**

Ans. It is a connecting lead embedded in the brush which maintains good electrical connection between the brushes and commutator and the external circuit.

WORKING PRINCIPLE

Q. **What is the basic working principle of the D.C. generator?**

Ans. Faraday's law of electromagnetic induction.

Q. **What is the working principle of D.C. generator?**

Ans. "Whenever a conductor cuts the magnetic flux an e.m.f. is induced in it." The induced e.m.f. is directly proportional to the rate of change of flux and the number of conductors.

Q. **If you increase the rate of change of flux will the induced e.m.f. increase or decrease?**

Ans. Increase.

Q. **If the number of conductors are decreased, how the e.m.f. induced is going to be effected?**

Ans. Will decrease.

Q. **What is the e.m.f. equation of the D.C. generator?**

Ans.
$$E = \frac{\phi \cdot Z \cdot N \cdot P}{60 \times a} \text{ volts}$$

where E = e.m.f. induced in volts, ϕ = flux in Wb, Z = number of armature conductors, N = speed in rpm, P = number of poles and a = number of parallel paths.

Q. **What is ϕ in the e.m.f. equation of the D.C. generator?**

Ans. ϕ is the flux produced by the field winding (the field winding can be series field, shunt field or both). The flux is $\phi \propto I_f Z_f$ where I_f is the current flowing through the field winding and Z_f is the number of conductors in the winding.

Q. **How I_f and E are related?**

Ans. Both are directly proportional to each other.

Q. **How the e.m.f. induced in D.C. generator is concerned to the parallel paths?**

Ans. Both are inversely proportional to each other. The e.m.f. generated depends upon the number of conductors. If the number of conductors in a parallel path are less the e.m.f. generated will be less (lap winding) and if the conductors per parallel path are more, the e.m.f. generated will be more (wave winding).

Q. **If the number of poles in a generator are reduced to half and parallel paths are doubled what will happen to its generated e.m.f. per parallel path?**

Ans. It will be one fourth of the initially generated e.m.f. ($E/4$).

Q. **What will be the current rating of the generator?**

Ans. It will be four times the previous one.

Q. **Can you tell what is the main factor which actually increased the current capacity four times?**

Ans. It is the number of parallel paths. Otherwise the current rating per parallel path is same. In parallel all the currents per parallel path are added to make it four times.

Q. **The D.C. armature winding is of open type or closed type?**

Ans. Closed type windings.

Q. **What types of windings are used in D.C. armature winding?**

Ans. Lap winding and wave winding.

Q. **What do you mean by lap winding?**

Ans. The winding in which the number of parallel paths are equal to the number of poles in the machine is known as lap winding.

Q. **What is wave winding?**

Ans. The winding which has got only two parallel paths whatsoever the number of poles may be, is known as wave winding.

Q. **Which winding is preferred for high voltage generation?**

Ans. Wave winding.

Q. **Can you use wave winding for high current, if no, name the winding?**

Ans. No, lap winding.

TYPES OF GENERATORS

Q. **What are the main two classifications of D.C. generator?**

Ans.
1. Permanent magnet type generator.
2. Electromagnet type generator.

Q. **What are the types of electromagnet type generators?**

Ans. These are separately excited and self-excited generator.

Q. **Can you classify the self-excited generator?**

Ans. Series generator, shunt generator and compound generator.

Q. **Why these generators are called as self-excited generator?**

Ans. Because in these generators the field winding is excited from the same generated voltage and extra excitor is not needed.

Q. **What is the main reason for inducing e.m.f. in the self-excited generator?**

Ans. Residual magnetism.

Q. **What is series generator?**

Ans. The generator in which the field winding is connected in series with the armature is known as series generator.

Q. **What is the resistance of series winding?**

Ans. Very low resistance say 0.3 Ω, 0.5 Ω.

Q. **How the series field winding is wound?**

Ans. It is wound with less number of turns of thick conductor, as to have low resistance.

Q. **Why it is desired to have low resistance of the series field winding?**

Ans. If the field resistance is high, the major part of the generated voltage will be consumed in the field winding and net voltage for external load will be less.

Q. **If you are starting a D.C. series generator what should be the position of the load switch?**

Ans. Should be closed, otherwise the e.m.f. will not be induced.

Q. **How will you define the shunt generator?**

Ans. The generator in which the field winding is connected across the armature is known as shunt generator.

Q. **What is supposed about the resistance of the field winding in case of the shunt generator?**

Ans. The shunt field resistance should be high enough to bear the generated voltage, as the field winding is connected across the armature.

Q. **How the shunt field winding is wound?**

Ans. It is wound with thin wire having more number of turns so that the resistance be more.

Q. **What is the value of shunt field current?**

Ans. It is less enough, generally it varies from 5 to 10% of the total load current for which the generator is designed.

Q. **Which generator will you prefer for battery charging?**

Ans. Shunt generator.

Q. **What should be the position of the load switch while the shunt generator is to be started?**

Ans. Should be positively opened.

Q. **Is the field current a depending factor on the load current in case of shunt generator?**

Ans. Negligible, practically no.

Q. **Name the generator in which the field current depends upon the load current?**

Ans. Series generator.

Q. **Which generator is generally used for the excitor in alternator etc.?**

Ans. Shunt generator.

Q. **If both series and shunt field windings are used in a generator, what is the name of the generator?**

Ans. D.C. compound generator.

Q. **How the D.C. compound generators are classified according to the shunt field connections?**

Ans.
1. Long shunt generator.
2. Short shunt generator.

Q. **The differential compound generator is that generator in which the current in field windings shunt and series is just opposing to each other, but if it helps each other what is the name given?**

Ans. Cumulative compound generator.

Q. **If a magnetic pole of a D.C. compound generator is taken out with winding intact, can you tell the type if (a) there is one winding of thick wire (b) there is one winding of thin wire (c) four terminals are brought out two thin and two thick?**

Ans. Yes, by seeing it can be stated as

(a) series generator

(b) shunt generator

(c) compound generator.

Q. **Which generator is used in long transmission line as a booster?**

Ans. Series generator.

Q. **Can you tell the three classifications of D.C. compound generator according to the number of turns in series winding?**

Ans. Level compound, over compound and under compound.

Q. **It is desired to supply a load at a very far distance, which compound generator will you you recommend?**

Ans. Over compound generator.

Q. **What do you mean by the armature drop?**

Ans. The voltage drop in the armature because of the load current
$V_a = I_a \times R_a$ volts.

Q. **Define short shunt compound generator?**

Ans. The compound generator in which the shunt field winding is directly shunted across the armature is known as short shunt compound generator.

Q. **What is magnetic drag?**

Ans. As we know that whenever a current carrying conductor is placed in the magnetic field it experiences a force and tends to rotate.

Q. **What do you mean by the back torque?**

Ans. In case of D.C. generators, whenever the load comes, the armature experiences a torque because of the load current in opposite direction of the rotor of prime mover. So it is called as back torque.

Q. **What are the effects of back torque?**

Ans. Due to back torque, the power of the prime mover is determined and kept more than the maximum output of the generator.

Q. **What foremost precautions should be observed while starting a D.C. generator?**

Ans. The shunt generator in no case be started with load switch ON. The load switch must be off.

Q. **Why is it necessary?**

Ans. It is because the e.m.f. induced due to residual magnetism is very small, and if the load switch is ON, the emf will have tendency to move the electrons to load side (having less resistance than field winding). As a result the field winding will not be magnetised to full strength as a result the voltage will not be generated anymore.

Q. **If you start a series generator with load switch opened, what will happened?**

Ans. The D.C. generator will not generate the voltage.

Q. **Why D.C. generator will not generate?**

Ans. Because the opened load switch will not allow the current to complete its path and field winding will not produce more flux than the residual one. So only voltage due to residual magnetism will be available.

Q. **Then which precaution be adopted in D.C. generator case?**

Ans. The load switch should be ON while starting.

Q. **Which is the easy way of increasing the load voltage while supplying load?**

Ans. Control the field winding current because $E \propto I_f$.

ARMATURE REACTION

Q. **What is the armature reaction?**

Ans. The armature reaction is defined as "the effect of the magnetic field produced by the armature conductors on the distribution of flux under the main poles."

Q. **What do you understand by the armature flux?**

Ans. Whenever the load is put on the generator, the armature conductor carries load current which produces the magnetic flux (Ampere's law). This flux is known as armature flux.

Q. **What are the effects of the armature reaction?**

Ans.
1. It weakens the main magnetic field.
2. It distorts the main magnetic field.
3. It changes the m.n.p. position.
4. Reduces the voltage induced.
5. Produces the sparking on the commutator.
6. The flux becomes strong on T.P.T. and weak on L.P.T.

Q. **In which direction the m.n.p. is shifted in case of D.C. generator?**

Ans. In forward direction, i.e., in the direction of rotation of the armature.

Q. **Does the angle of lead and load current have any relationship?**

Ans. Yes, the angle depends upon load; more load current more angle and less load current less angle.

Q. **In case of forward lead, the armature flux is inclined thus the flux produced by the armature has two components, one is cross magnetising field, what is the name of the other component?**

Ans. Demagnetising field.

Q. **If by any means, the cross magnetising field is made zero, will there be any necessity for shifting the brush position for correct m.n.p. position?**

Ans. Not at all.

Q. **In case of loading the generator, the flux distribution on the pole face will be uniform or not?**

Ans. No, it will be strong at T.P.T. and weak at L.P.T. in case of generator.

Q. **What do you mean by the demagnetising flux?**

Ans. The armature flux which is directly opposing the main magnetic field is known as the demagnetising flux.

Q. **How the armature reaction is minimised?**

Ans.
1. By using interpoles.
2. By giving forward lead to the brushes.
3. By using compensating winding.

Q. **What is an interpole?**

Ans. It is a magnetic pole situated at MNA (between two main poles) and has approximately half or one-third size of the main pole.

Q. **What is the main function of the interpole?**

Ans. To improve armature reaction and to minimise the sparking.

Q. **How the polarity is kept?**

Ans. It is same as that of the polarity of the main pole ahead in the direction of rotation of the armature.

Q. **What actually the interpole does?**

Ans. It neutralizes the cross magnetising flux produced by the armature.

Q. **How the demagnetising effect is minimised?**

Ans. By increasing the series field turns.

Q. **Are the demagnetising and cross-magnetising fluxes constant for a particular generator?**

Ans. No.

Q. **On what factor does it depend?**

Ans. Load current.

Q. **When the generator is not supplying, will it be having ϕ_a, and ϕ_d components?**

Ans. Practically no.

Q. **But the generator is supplying the shunt field winding of the generator?**

Ans. The shunt field current is very less in comparison to the full load current, so it is neglected.

Q. **How the connections are made for interpole winding?**

Ans. It is in series with the armature, so that the flux produced by the interpole corresponds to the load current.

COMMUTATION

Q. **What do you mean by the commutation?**

Ans. Commutation means the changes taking place in an armature coil during the period of short circuit by the brushes when it goes under opposite poles.

Q. **What is the reason of sparking on the commutator?**

Ans. The main reason of sparking is the failure of the current to reach its full value in opposite direction, in the coil undergoing commutation.

Q. **Why the current fails to reach to its full value in opposite direction?**

Ans. Because of the reactance voltage.

Q. **What do you understand by the term reactance voltage in commutation?**

Ans. When the current rises from zero to full load current value in opposite direction, this rising current produces some magnetic flux and hence voltage in the coil due to self-induction. According to Lenz's law, this reactance voltage will oppose the rising current and cause some delay, resulting in sparking at commutator.

Q. **How the reactance voltage is calculated?**

Ans. Reactance voltage $= L \times \frac{2I}{t}$

where L = inductance in henry, I = current in the coil, and t = time of commutation.

Q. **What are the effects of commutation?**

Ans.
1. Sparking at commutator.
2. Life of commutator reduces because of sparking.
3. Less output of the generator.
4. More temperature rise in armature winding.

Q. **The sparking at commutator depends upon which factor?**

Ans. Load current.

Q. **Where the compensating winding is housed?**

Ans. In the pole faces.

Q. **How you can improve the commutation?**

Ans.
1. By using compensating winding or interpoles.
2. By using high resistance brushes.
3. By giving a slight forward lead to the brushes.

Q. **How the tension on the brushes is exerted?**

Ans. By means of a spring mounted on the brushes holder (rocker). It has a mechanical arrangement of adjustment.

PARALLEL OPERATION

Q. **What do you mean by parallel operation?**

Ans. To run more than one generator in parallel for meeting the load demand is known as parallel operation.

Q. **What is the necessity of parallel operation?**

Ans.
1. To meet the actual load demand.
2. To facilitate the overhauling of generators.
3. Continuation of electric supply in case of faulty generator.
4. Maximum efficiency of operation.

Q. **What are busbar and feeder?**

Ans. The busbars are generally copper bars flat in shape and run behind the switchboard in a box. In D.C. these are positive and negative and sometimes the equalizing busbar also.

The feeders are the wires connected to the busbar. These deliver supply to different areas.

Q. **What are the conditions to be fulfilled before parallel operation?**

Ans.
1. The polarity should be same.
2. The voltage of busbar and incoming generator should be equal.

Q. **How the voltage is checked?**

Ans. By means of a voltmeter.

Q. **Can you check the polarity by means of the voltmeter?**

Ans. Yes, by means of a center zero voltmeter.

Q. **How the shifting of load is done?**

Ans. Just by increasing the field excitation of the generator to be loaded.

Q. **To isolate the generator from the line, should the prime mover be stopped first or the circuit breaker be put off first?**

Ans. The circuit breaker should be off that too after reducing the load current to zero amperes.

Q. **If we put the prime mover off, what will happen?**

Ans. There is every possibility that this generator will run as a D.C. motor and hence the prime mover causing extra load on the busbar.

Q. **Should the generator be put off when it carries a load?**

Ans. No, first the current delivered be brought to zero value and then put off the main switch or circuit breaker.

Q. **How many busbars are used in case of compound generator?**

Ans. Three—the positive, negative and equalizing busbars.

Q. **What is the function of the equalizing busbar?**

Ans. It is used to share the load on the compound generators. It also helps to load a generator and automatic distribution of load in running condition.

Q. **What is the difference between equalizing busbar and equalizer?**

Ans. Both are same.

Q. **Is it possible to share the load on compound generators without equalizing busbar?**

Ans. No.

Q. **What are the characteristics of the D.C. generator?**

Ans. These are :

1. No load saturation characteristic.
2. Internal or total characteristic.
3. External characteristic.

Q. **What is no load characteristic?**

Ans. It shows the relationship between the no load generated e.m.f. and field current say excitation.

Q. **What do you understand by the external characteristic of the D.C. generator?**

Ans. It shows the relationship between the load current and terminal voltage.

Q. **Name the characteristic which states the relationship between the e.m.f. actually generated and armature current?**

Ans. Internal or total characteristic.

Q. **Which type of D.C. compound generator is used for welding purpose?**

Ans. Differential compound generator.

Q. **What is meant by critical resistance of a D.C. series generator?**

Ans. It is the maximum resistance with which the D.C. generator will be able to excite.

EFFICIENCY

Q. **What do you understand by the term efficiency?**

Ans. It is the ratio of the output to input generally shown in percentage.

Q. **What is the output of a generator?**

Ans. It is the electrical energy and is the product of voltage and current in watts or kW.

Q. **What is the input to a generator?**

Ans. It is the mechanical energy generally expressed in horsepower.

Q. **Can a generator by designed for 100% efficiency?**

Ans. No.

Q. **Why output is not equal to input?**

Ans. Because of losses—the copper losses, iron and friction losses, winding losses, etc.

Q. **What are copper losses?**

Ans. The power losses in copper portion of the generator are known as the copper losses. These are shunt field copper losses, armature copper losses, series field copper losses and interpole copper losses, etc.

Q. **Which losses are constant losses and remain constant from no load to full load?**

Ans. Iron and friction losses.

Q. **Why copper losses are not constant losses?**

Ans. Copper losses are proportional to the square of current hence these change with the increasing or decreasing of load current.

Q. **How the armature and shunt field copper losses are calculated?**

Ans. Armature copper losses $W_a = I_a^2 \times R_a$ watts

Shunt field copper losses $W_{sh} = I_{sh}^2 \times R_{sh}$ watts

where I_a and I_{sh} are currents in armature and shunt field windings, R_a and R_{sh} the resistances of armature and shunt field windings, respectively.

Q. **Why a generator fails to build up the voltage?**

Ans. It can be because of these reasons :

1. Loss of residual magnetism.
2. Reverse d.o.r.
3. Condition of load (more and more load).
4. Brushes not in m.n.p. position.
5. Open circuit in field or armature circuit.

Q. **Why a generator is overheated?**

Ans. It can be because of :

1. Sparking on the commutator.
2. Defective bearings.
3. Short circuit in field coil or armature.

OBJECTIVE TYPE MULTIPLE CHOICE QUESTIONS

1. The D.C. generator changes the mechanical energy into ......
   (a) chemical energy (b) electrical energy
   (c) D.C. electrical energy (d) solar energy
2. The electrical energy initially induced in a conductor is of ...... nature.
   (a) D.C. electrical (b) A.C. electrical
   (c) pulsating D.C. (d) None of the above.
3. In a two pole D.C. generator, the complete revolution of a particular conductor induces ......

(a) two cycle of A.C. (b) one cycle of A.C.
(c) four cycles of A.C. (d) half cycle of A.C.

4. The initially induced e.m.f. when taken out as D.C. the ...... is provided on the shaft.
(a) commutator (b) slipring
(c) end ring (d) None of the above.

5. The brushes are generally provided on the ......
(a) magnetic neutral axis (b) magnetic axis
(c) shaft anywhere (d) shaft in the centre

6. The assembly where brushes are provided for the collection of e.m.f. is known as ......
(a) slipring (b) commutator
(c) rocker (d) interpole

7. The brushes are held with pressure in the brush holder with the help of ......
(a) pig tail (b) spring (spring tension)
(c) screws (d) adjustable spring loaded lever.

8. The commutator fitted on the shaft is ......
(a) insulated from the shaft
(b) insulated from the armature stampings
(c) insulated from the all metallic constructions
(d) All the above.

9. The pole windings are housed on the ......
(a) poles (b) armature (c) rocker (d) shaft

10. The interpoles are provided ......
(a) in between the main poles (b) on the shaft
(c) all along the main poles (d) only on the top of the body

11. The direction of the voltage induced in an armature of a D.C. generator is determined by using ......
(a) Fleming's right hand rule (b) Fleming's left hand rule
(c) screw rule (d) Ampere's law.

12. All D.C. machines are characterised by ......
(a) flux (b) brushes
(c) parallel paths (d) commutator.

13. For higher voltage ...... winding is used.
(a) lap (b) wave
(c) compensating (d) equalizing.

14. Lap winding is used in case when ......
(a) more voltage and less current is required.

(b) more current and more voltage is required.
(c) less voltage and more current is required.
(d) less voltage and less current is required.

15. The e.m.f. equation of the D.C. generator is ......
(a) $E = 2.22k_f \cdot k_d \cdot k_z \cdot \phi_m \cdot Z \cdot f$ volts
(b) $E = \frac{\phi \cdot Z \cdot N \cdot P}{60 \cdot a}$ volts
(c) $E = 4.44\phi_m \cdot f \cdot N$ volts
(d) $E = 3.44B_m \cdot A \cdot f \cdot N$ volts.

16. The 'a' in e.m.f. equation of the D.C. generator stands for ......
(a) the number of brushes in the machine.
(b) the number of parallel paths in the winding.
(c) the number of conductors in a parallel path.
(d) the number of armature coils.

17. If the flux of a D.C. generator is doubled the e.m.f. induced will be ......
(a) halved (b) doubled (c) zero volt (d) same

18. Keeping every thing unchanged, the speed of the prime mover will have ...... effect.
(a) proportionality (b) reciprocative
(c) no (d) None of the above.

19. If the speed of a D.C. generator is 500 rpm and the speed is accelerated to 1000 rpm, keeping other factors constant, the induced e.m.f. will be ......
(a) doubled (b) halved
(c) same (d) one and half time

20. If the amount of flux is reduced to half, the speed of the prime mover is doubled, the e.m.f. induced will be ......
(a) halved (b) doubled (c) one forth (d) same

21. If an armature which was wound for two pole lap winding to now rewound for two pole wave winding the induced e.m.f. will be ......
(a) doubled (b) halved (c) same (d) zero

22. A D.C. shunt generator without residual magnetism is running at 400 rpm. Now keeping everything unchanged the speed is increased to 800 rpm, the emf induced will now be ......
(a) doubled (b) halved (c) zero (d) maximum.

23. The e.m.f. induced due to residual magnetism is in the order of ......
(a) 5-15 V (b) 30-50 V (c) 100-150 V (d) 400-450 V.

24. In series generator, the armature and field windings are connected in ......
(a) parallel (b) series

(c) shunted (d) series-parallel

25. The shunt generator is used for ......
    (a) battery charging (b) trains and trams
    (c) traction supply (d) amplification
26. The compound generator has both ...... windings.
    (a) series and complex (b) series and shunt
    (c) interpole and compensatory (d) series and interpole.
27. The pole winding which is connected in series with the armature winding is called ......
    (a) interpole winding (b) shunt field winding
    (c) auxiliary winding (d) damping winding.
28. The number of brushes in wave winding are ......
    (a) two whatsoever the number of poles may be.
    (b) four whatsoever the number of poles may be.
    (c) equal to number of poles.
    (d) equal to the number of parallel paths.
29. In D.C. generator, the total percentage of copper losses are ......
    (a) 5 to 10% of the full load losses.
    (b) 30 to 40% of the full load losses.
    (c) 0 to 50% of the full load losses.
    (d) 50 to 60% of the full load losses.
30. If the thickness of the laminations of a core is halved the eddy current losses becomes ......
    (a) doubled (b) halved
    (c) remains same (d) one-fourth
31. The brushes are always placed along ......
    (a) magnetic neutral axis (b) magnetic axis
    (c) geometrical neutral axis (d) in between GNA and MNA.
32. The failure of current in reaching to its maximum value in opposite direction while undergoing commutation is because of ......
    (a) the reactance voltage
    (b) the air gap between commutator and brushes
    (c) the north pole of field winding
    (d) more current.
33. A D.C. shunt generator is always started with ......
    (a) load switch on (b) load switch off
    (c) in any of the 'ON' or 'OFF' conditions
    (d) short-circuited armature.

34. The D.C. series generator is always started with ......
    (a) load switch 'ON'
    (b) load switch 'OFF'
    (c) in any of the position of 'ON' or 'OFF'.
    (d) short-circuited series field winding.

35. The generated voltage of a D.C. shunt generator may be increased ......
    (a) by increasing the speed of the prime mover.
    (b) by increasing the speed or by increasing the field current
    (c) by increasing the speed or by decreasing the field current.
    (d) None of the above.

36. The back torque in D.C. generators is directly proportional to the ......
    (a) field current
    (b) armature current
    (c) load current
    (d) series field current

37. The back torque actually decides the capacity of the ......
    (a) prime mover
    (b) load current
    (c) shunt field current
    (d) live current

38. The back torque is always ...... the prime torque produced by prime mover.
    (a) more than
    (b) less than
    (c) equal to
    (d) not related to

39. A generator will have minimum (approximately zero) back torque when it is ......
    (a) supplying full load current
    (b) supplying half load current
    (c) supplying zero ampere load current
    (d) supplying 10% load current

40. A D.C. shunt generator will produce maximum back torque when supplying ...... load current.
    (a) full
    (b) half
    (c) nil
    (d) quarter

41. The shunt field copper losses are proportional to the square of ...... multiplied by ......
    (a) armature current, voltage
    (b) shunt field current, shunt field resistance
    (c) field current, square of voltage
    (d) armature current, rotor resistance

42. The copper losses of the generator core are ......
    (a) armature copper losses
    (b) shunt field copper losses
    (c) series field copper losses
    (d) All the above.

43. The copper losses can be termed as ......
    (a) constant losses
    (b) variable losses
    (c) stray losses
    (d) 5% of the input

44. The constant losses in D.C. generator are ......
    (a) Eddy current and hysteresis losses
    (b) shunt field losses
    (c) windage losses
    (d) All the above.

45. A D.C. generator will have maximum efficiency if the variable losses are ...... the constant losses.
    (a) more than
    (b) less than
    (c) equal to
    (d) not related to

46. The effect of the flux produced by the armature conductors on the distribution of flux under the main poles is termed as ......
    (a) commutation
    (b) back torque
    (c) back e.m.f.
    (d) armature reaction

47. The effects of the armature reaction is ......
    (a) the main magnetic field distortion
    (b) TPT are strongly magnetised and LPT weakly
    (c) the MNP is shifted
    (d) All the above.

48. In case of D.C. shunt generator to bring the brushes at MNP, the brushes are given ......
    (a) forward lead in the direction of rotation of armature
    (b) backward lead to the direction of rotation of the armature
    (c) no movement
    (d) None of the above.

49. The changes a coil experiences while undergoing from one pole to another pole (passing through MNP) are studied under ......
    (a) armature reaction
    (b) interpoles
    (c) commutation
    (d) parallel operation

50. The poles which are small in size, placed in between the main poles are known as ......
    (a) series poles
    (b) shunt poles
    (c) interpoles
    (d) compound poles

51. The polarity of an interpole in case of D.C. generator is ......
    (a) always north pole on the top side of the generator.
    (b) always south pole on the top side of the generator.
    (c) always same as that of main pole ahead in the direction of rotation of the armature.
    (d) always opposite to that of the main pole ahead in the direction of rotation of the armature.

52. The interpole's winding is done with ......
    (a) thick copper wire
    (b) thin copper wire

(c) eureka wire (d) manganin wire

53. The interpole winding is connected in ......
    (a) series of the armature
    (b) in parallel of the armature
    (c) either in series or in parallel of the armature
    (d) nowhere in the armature circuit

54. Commutating poles are used to improve ......
    (a) armature reaction (b) commutation
    (c) parallel operation (d) series winding drops

55. The direction of rotation of a D.C. generator can be reversed by interchanging ......
    (a) the supply terminals
    (b) the armature connections
    (c) the field winding connections
    (d) the direction of rotation of the prime mover

56. If there is no residual magnetism the ...... will not be effected and will generate voltage.
    (a) separately excited generator (b) shunt field generator
    (c) series field generator (d) compound field generator

57. The armature reaction can be improved by
    (a) giving the brushes a forward lead in the direction of rotation of the armature.
    (b) using interpoles in the machine.
    (c) using compensating winding in the poles shoes.
    (d) All the above.

58. The forward angle of lead to the brushes in the direction of rotation of the armature depends upon ......
    (a) physical power of the operation
    (b) the space available
    (c) the amount of load
    (d) the size of the machine

59. The commutation can be improved by ......
    (a) using compensating or interpoles windings
    (b) using high resistance brushes
    (c) giving slightly forward lead to the brush rocker
    (d) All the above.

60. The commutation causes ......
    (a) sparking at commutator
    (b) short life of commutator

(c) less output of generator and more temperature rise
(d) All the above stated.

61. $W \propto B^{1.6} \cdot f \cdot V$ watts, where B is the flux density, f frequency in Hz and V is ...
(a) volume of the iron part
(b) voltage generated in armature
(c) voltage given to the windings
(d) circular velocity of the armature.

62. The maximum torque developed in D.C. machine is possible when ......
(a) the back e.m.f. = supply voltage
(b) the back e.m.f. = $\frac{1}{2} \times$ the supply voltage
(c) the back e.m.f. = $\frac{3}{4} \times$ the supply voltage
(d) the back e.m.f. = $2 \times$ the supply voltage

63. The torque developed by the D.C. motor is maximum when the ratio of applied voltage to back emf is ......
(a) 0.5 (b) 2 (c) 0.25 (d) 4

64. In a D.C. shunt motor case if the load on the motor is doubled and the applied voltage is halved, will the armature current drawn be .... times.
(a) 2 (b) 6 (c) 0.5 (d) same

65. State true or false.
(a) D.C. generator works on the principle of Lenz's law.
(b) Fleming's right hand rule is used to find out the direction of induced e.m.f.
(c) D.C. generator converts chemical energy into electrical energy.
(d) Self-excited generator requires a separate source of field excitation.
(e) Residual magnetism is essential for all self-excited generators.
(f) Series generator should not be started without load.
(g) Shunt generator is always started with load.
(h) Shunt generator is used for variable load.
(i) Differential compound generator are used for battery charging.
(j) Shunt generator is recommended for welding purpose.

66. Which statement is correct?
(a) Voltage regulation of over compounded generator is negative.
(b) The brushes are given backward lead for m.n.p. position in case of D.C. generator.

(c) The air gap between the interpoles and armature cores is less than the air gap between main poles and armature cores.

(d) Interpoles are connected in series with the shunt field winding.

(e) Shunt field winding is wound with less number of turns of thick wire.

(f) Series field winding is wound with less number of turns of thick wire.

(g) Series generator is used as a feeder booster.

(h) Eddy current losses in D.C. generator are minimised by using laminated armature.

(i) Hysteresis losses can be minimised by using soft iron cores.

(j) Equalizing busbar is used for parallel operation of D.C. compound generator.

67. Fill in the blanks.

(a) The e.m.f. induced in D.C. generator depends upon ......

(b) The winding which is embedded in the pole face of D.C. generator is known as ......

(c) In D.C. machines ...... is used to hold brushes.

(d) Pigtail is attached with ......

(e) Brushes are used to ...... the e.m.f. from commutator.

(f) The laminations are ...... from each other.

(g) The segment of commutator are insulated from each other by ......

(h) In D.C. cumulative compound generator the flux of shunt field winding and series field winding ...... each other.

(i) Armature winding connections are ...... to the commutator segment.

(j) Armature windings are of two types the lap and ......

68. The output of D.C. machines is proportional to the product ......

(a) D^2L (b) DL^2 (c) DL (d) D^3L

(where D is the diameter of armature and L is the length of core).

69. The interpole winding is connected in series with the ......

(a) armature (b) field coils

(c) shunt field winding (d) None of the above.

70. The assembly of number of copper bars or segments separated from one another by a suitable insulating material and connected to armature conductors is known as ......

(a) armature cores (b) rotor bars

(c) compensating winding (d) commutator

71. The effect of armature reaction is ......

(a) reduction in e.m.f.

(b) increase in iron losses

(c) sparking at commutator

(d) reduction in e.m.f. increase in iron losses, sparking at commutator and delayed commutation.

72. The height of the pole is decided by the ...... to be provided on the pole at full load ......

(a) m.m.f. (b) e.m.f.

(c) reluctance (d) flux density.

73. In D.C. shunt generator, the eddy current losses ......

(a) vary inversely as the square of the lamination thickness.

(b) vary inversely as the square of the flux density.

(c) occur in every conductive material.

(d) occur only in armature.

74. State true or false.

(a) The D.C. shunt generator should always be started with full load.

(b) The series generator should be started with load.

(c) The on load voltage of D.C. shunt generator is more than the no load voltage.

(d) The armature reaction distorts the main magnetic field.

(e) The residual magnetism is the backbone of self-excited generator.

CHAPTER 14

D.C. Motor

INTRODUCTION

Q. **What is D.C. motor?**

Ans. The machine which converts D.C. electrical energy into mechanical energy is known as D.C. motor.

Q. **Where actually the mechanical energy is available?**

Ans. On pulley.

CONSTRUCTION

Q. **What are the parts of a D.C. motor?**

Ans. The motor has two types of parts—the rotary and stationary parts. The rotary parts are those which revolve during the working of motor and stationary parts are those which do not revolve rather remain stationary even during the working of the motor.

Q. **Name some stationary parts of D.C. motor?**

Ans. These are body or yoke, poles, bed sheet, terminal box, rocker, side covers etc.

Q. **What are rotary parts?**

Ans. The parts are armature, shaft, pulley, bearings, commutator, fan etc.

Q. **The frame of the D.C. motor is generally made of which material?**

Ans. Cast iron.

Q. **The commutator is made up of which substance?**

Ans. Mainly it is made of copper segments.

Q. **How actually the commutator is assembled?**

Ans. It is actually assembled of the copper segments. The segments are insulated from each other by means of mica. The segments are tapered so that when all are assembled a round shape may be obtained.

Q. **What is the main function of the commutator?**

Ans. It actually helps in converting the initially induced A.C. emf into D.C. for the external load.

Q. **How does it perform?**

Ans. With the help of rocker and brushes, as both remain stationary and collect emf.

Q. **What is rocker?**

Ans. The rocker is an assembly which has brush holders housings. Generally it is round in shape fixed at commutator side with a provision of moving forward and backward depending upon the requirement.

Q. **How the brushes are fitted in the housings?**

Ans. The rocker has the rectangular (generally) housings. A spring loaded strip provides spring tension so as to have required necessary pressure on the brush to be in contact with the commutator. The pig tail of brush can be screwed also.

Q. **What is the function of the fan?**

Ans. The fan circulates the air through the machine as to provide cooling.

Q. **A D.C. motor which has only two poles is known as ......**

Ans. A bipolar machine.

Q. **Where actually the emf is initially induced?**

Ans. It is induced in the armature.

Q. **The D.C. motor armature is made of lamination of silicone steel or simply steel laminations?**

Ans. Lamination of silicone steel highly compressed and riveted, so as not to leave any air space between them.

Q. **Are these laminations short-circuited together?**

Ans. No, in spite of forming a solid enblock these are not behaving like a solid one. These laminations are insulated from each other with the help of a thin layer of insulating varnish.

Q. **When the laminations are assembled, apart from the slots these are having other holes, which after assembly form a duct-shape. What do you call these ducts?**

Ans. These are ventilating ducts and are used to provide air circulation throughout the cores dissipating the heat developed.

Q. **Are these ducts only horizontal or radial?**

Ans. No, sometimes in big machines both horizontal and radial ducts are provided because the heat developed is more. In small machine only horizontal ducts are provided.

Q. **The commutator is generally made from hard drawn copper, but nowadays for modern D.C. machines which material is used?**

Ans. Silvered copper.

Q. **Can you have any idea about the thickness of mica or micanite sheet used in between the segments of commutator?**

Ans. Generally 0.8 mm thick sheet of mica or micanite is used.

Q. **What are the types or fashions in which brushes are applied on the commutator?**

Ans. The brushes applied on the commutator may be of three types—the radial, trailing and reaction types.

Q. **Can you tell the difference?**

Ans. The one having brushes set so that their line is radial to the commutator are **radial type**. The other in which brushes are inclined in a trailing direction are **trailing type** and the brushes fashioned inclined in leading direction are known as **reaction type**.

Q. **How the copper losses are calculated?**

Ans. These are equal to the product of square of current multiplied by resistance.

Q. **If a D.C. machine field winding is fed from D.C. source in one case from a D.C. generator in other case from a rectifier, will the copper losses be same in both cases?**

Ans. Now actually the copper losses are also proportional to the form factor of the current. In case when supplied by D.C. generator the form factor is unity so the losses will be only I^2R but in case when the current is supplied from rectifier, the form factor is not unity but much higher than unity, so the losses will be much higher.

Q. **Which motor needs more heat dissipating source the one fed from D.C. generator or fed from rectifier unit?**

Ans. The one fed from rectifier unit.

WORKING PRINCIPLE

Q. **What is the working principle of a D.C. motor?**

Ans. Whenever a current carrying conductor is placed in the magnetic field, a torque is produced over the conductor which tends to rotate it.

Q. **Where the torque is maximum?**

Ans. At right angle to the magnetic field.

Q. **How the direction of torque is determined?**

Ans. By means of Fleming's left hand rule.

Q. **What is Fleming's left hand rule?**

Ans. Stretch your left hand in such a way so that the thumb, middle finger and forefinger are at right angles to each other. Now if the first finger indicates the direction of flux, middle finger the direction of current in the conductor, then the thumb will indicate the direction of motion or the torque developed.

Q. **How the d.o.r. of a D.C. motor can be changed?**

Ans. By changing the direction of current in either the field or the armature.

Q. **Why it is customary to change the direction of current in armature?**

Ans. It is easy and changing of direction of current in the field winding may neutralize the residual magnetism.

Q. **If the direction of current in both field and armature is changed, will the d.o.r. of the motor change?**

Ans. No, it will be same as it was.

BACK E.M.F.

Q. **What is the back e.m.f.?**

Ans. The e.m.f. induced in armature of a D.C. motor while running is called back e.m.f.

Q. **What is the significance of the back e.m.f.?**

Ans. Whenever the armature of a D.C. motor is rotated in the magnetic field as a result the e.m.f. is induced in the armature, this induced e.m.f. will oppose the cause responsible for the production of this e.m.f. i.e. the voltage fed. This is the reason we call it back e.m.f. or counter e.m.f. i.e. E_b.

Q. **Can you calculate the back e.m.f. If yes, what is the formula?**

Ans. Yes, $E_b = \dfrac{\phi \cdot Z \cdot N \cdot P}{60 \cdot a}$ volts

Q. **What is the relationship between E_b and V?**

Ans. $E_b = V - I_a R_a$.

Q. **What is the power equation of the motor?**

Ans. $E_b \cdot I_a = V_a \cdot I_a - I_a^2 \cdot R_a$

Q. **What is the condition for maximum power?**

Ans. $E_b = \dfrac{V}{2}$

Q. **What is the back e.m.f. at standstill condition of the rotor?**

Ans. Zero.

Q. **Can you calculate the armature current if the back e.m.f. and armature resistance are known?**

Ans. Yes, $I_a = \dfrac{V - E_b}{R_a}$ amp.

SPEED AND SPEED CONTROL

Q. **On what factors the speed of a D.C. motor depends?**

Ans. The speed depends upon the flux, the poles, the voltage applied and the armature circuit resistance.

Q. **Does the speed have any relationship between the back e.m.f. and flux?**

Ans. Yes, $N = K\frac{E_b}{\phi}$

where N = speed in rpm, k constant, E_b = back e.m.f. and ϕ the flux.

Q. **On what factors the constant *k* depends?**

Ans. It depends upon the number of armature conductors, type of winding, number of poles and parallel paths in the winding.

Q. **What are the methods for controlling the speed?**

Ans.
1. The field control.
2. The armature control.
3. The voltage control.
4. The Ward Leonard control.

Q. **What is the armature control?**

Ans. In this method, the voltage fed to the armature circuit is controlled by connecting the variable resistance in the armature circuit.

Q. **Can you tell the nature of the resistance connected in armature circuit?**

Ans. It is a resistance of low value and high current capacity.

Q. **The speed is controlled below or above normal?**

Ans. Below normal.

Q. **What is field control?**

Ans. In this method of speed controlling the variable resistance is inserted in the field winding, so that the flux produced by the field winding is controlled. We know that the speed is inversely proportional to the flux.

Q. **What type of resistance is placed in the field circuit?**

Ans. The resistance of low current capacity and high ohmic value.

Q. **Can you control the speed above normal?**

Ans. Yes, in this method speed is controlled above normal.

Q. **Whenever the voltage given to the motor is controlled what is the name given to this controlling?**

Ans. Voltage control method.

Q. **In voltage control method, how the speed is controlled?**

Ans. Below method.

Q. **What do you understand by the Ward Leonard method of speed controlling?**

Ans. In this method, the main motor is connected to the variable voltage (MG set). The voltage generated is controlled and the field winding of main motor is connected with reversible switch so both the methods of speed control i.e. the voltage control and field control are involved and even the change of d.o.r. of the motor is obtained.

Q. **How the armature control is obtained in case of series motor?**

Ans. By providing an armature diverter.

Q. **What do you know about the armature diverter?**

Ans. It is a variable resistance of high current capacity and low ohmic value, connected across the armature.

Q. **Can you apply the field control method by connecting a variable resistance in series with the field winding?**

Ans. No, in that case too a field diverter of low resistance and high current capacity is connected in parallel with the field winding.

Q. **Is there any difference between the resistances used for field control, in series motor and shunt motor?**

Ans. Yes, the resistance used for shunt motor is of high value and low current capacity but in case of series motor it is of low value and high current capacity.

Q. **Can the speed of a D.C. series motor be controlled by voltage control method?**

Ans. Yes.

Q. **Can you use the field tapped control directly in series motor for controlling the speed?**

Ans. Yes, in this method, the different tappings are taken out from the field winding directly to the rotating switch and flux variation can be obtained.

STARTING AND STARTERS

Q. **Can you connect a direct current motor directly to the line?**

Ans. No.

Q. **Name the appliance generally used for starting the D.C. motor?**

Ans. D.C. motor starter.

Q. **Why the D.C. motor should not be started directly from mains?**

Ans. Because at the time of starting, there is no back e.m.f. in the armature. The armature circuit has low resistance, so if the full line voltage is impressed, the armature current will be high enough to damage the winding of the motor.

Q. **What is a starter?**

Ans. The device used for starting the D.C. motor?

Ans. **What are the main functions of starter?**

Ans.
1. To protect the motor from taking overload current.
2. To protect the motor against supply failure.
3. To facilitate the starting and stopping of the motor.

Q. **How the supply failure is protected?**

Ans. The no volt release coil or holding magnet will de-energised and the spring will bring the handle to off position.

Q. **What is NVRC and how it is wound?**

Ans. It is the no volt release coil, connected across the supply through shunt field winding (3 point starters). It is wound with thin conductors having more number of turns.

Q. **Is it always connected to the mains through shunt field winding?**

Ans. No, in case of four point starter it has an independent circuit.

Q. **How overload protection is achieved?**

Ans. By providing the OLRC.

Q. **What is the OLRC and how it is connected and why?**

Ans. It is the overload release coil. It is connected in series with the motor so that motor current may flow through this coil and the protection be achieved accordingly.

Q. **How ORLC is wound?**

Ans. It is wound with thick conductor having less number of turns.

Q. **What type of overload relay is used in D.C. motor starters?**

Ans. Magnetic relays.

Q. **How the overload relay protects the motor from taking overload current?**

Ans. In normal working conditions, the magnetic flux produced by the overload relay is not sufficient to pull the strip and whenever overload comes the flux produced is more, the strip will be attracted, short circuiting both the terminal of the no volt release coil. It will result in the demagnetisation of the no volt release coil and the spring will pull the handle to the off position protecting the motor.

Q. **Is there any arrangement provided for setting the overload current?**

Ans. Yes an overload current setting is provided.

Q. **How the overload current is set?**

Ans. It depends upon the distance (space) between the electromagnet and the strip to be pulled, more distance more current, less distance less current.

Q. **Name the starters used for D.C. shunt and compound motors?**

Ans. Three point and four point D.C. motor starter.

Q. **Why a three point starter is called so?**

Ans. Because, it has only three terminals for external connections i.e. L, A and Z.

Q. **How many terminals are there in a four point starter?**

Ans. Four—L, N, A and Z.

Q. **What do you mean by L.A.N.Z. terminals?**

Ans. L—line, A—armature, N—negative and Z—shunt field winding.

Q. **Why negative terminal is brought out in a four point starter?**

Ans. To make the independent circuit of the no volt release coil.

Q. **Why it is desired to make an independent circuit of the no volt release coil?**

Ans. In case of shunt or compound motors whenever field control method of speed controlling is used, the external resistance is added, in field circuit which will decrease the current through the field circuit, reducing the flux produced by the no volt release coil and the pull will be certainly weakened to hold the handle. The spring attached will pull the handle to off position.

So to make the no volt release coil circuit independent of the field circuit current, the negative terminal is brought.

Q. **In case of D.C. four point starter, whenever the terminals of the no volt release coil are short circuited by means of the short circuiting strip, the positive and negative both are short circuited directly, will it not damage the starter?**

Ans. No, the positive and negative are not short circuited directly because a resistance of suitable value is placed in series to limit the current.

Q. **Can you recommend a D.C. three point starter for starting the D.C. series motor?**

Ans. No.

Q. **What are the common types of starter used for D.C. series motor?**

Ans.
1. The drum controller.
2. Face plate starter.
3. The liquid starter.

Q. **What do you understand by the drum controller for D.C. series motor?**

Ans. It is of the shape of a drum, there are a number of strips and fingers. The fingers are mounted on the insulated rod and are connected with the standard resistance. The strips are fixed over the drum, whenever the drum rotates, the finger makes connection with the resistance and actually helps in adding and removing the external resistance in the motor circuit. By this starter we can change the speed and the d.o.r. of the motor also.

Q. **Is there any protection for quenching the spark?**

Ans. Yes, a blowout coil is provided.

Q. **Is it advisable to start a D.C. series motor without load?**

Ans. No, it should be started with load.

Q. **Why D.C. series motor should not be started without load?**

Ans. At no load, the speed will be very high and it may damage the foundation even.

Q. **What are the face plate starters?**

Ans. These are having variable resistance to add. The no load protection and sometimes the overload protection is provided with these starters.

SPEED REGULATION

Q. **What do you understand by speed regulation.**

Ans. The speed of the D.C. motor decreases as the load on the motor is increased, this speed can be regulated by the speed control methods and the regulation is termed as speed regulation.

Q. **Define speed regulation in D.C. motors?**

Ans. The speed regulation is defined as the rise in speed, shown in the percentage of full load speed, when full load is thrown off, i.e.,

$$\% \text{ regulation} = \frac{N_0 - N_f}{N_f} \times 100$$

Q. **What do you mean by the no load speed and full load speed?**

Ans. The speed of the motor when it is not carrying any load is known as no load speed and if it carries full load the speed is known as full load speed.

Q. **How the speed is measured?**

Ans. By means of tachometer or speedometer with stop watch.

Q. **Can you calculate the speed regulation at half load?**

Ans. Yes, load the machine only half the full load, throw the switch off, see the rise in speed and calculate.

TESTING OF D.C. MACHINE

Q. **What is the necessity of testing the D.C. machine?**

Ans. To be acquainted with the performance of the machine at different loads efficiency, temperature rise, etc.

Q. **What are the different methods of testing the D.C. machine?**

Ans.
1. Direct testing method.
2. Indirect testing method.
3. Hopkinson's or Regenerative test.

Q. **What is meant by direct testing?**

Ans. The testing method in which actually the machine is loaded and every winding and operations undergo all normal loading conditions.

Q. **What is indirect testing?**

Ans. The test in which machine is not actually loaded but all the losses and efficiency are calculated by means of the values of current and resistance etc., so obtained.

Q. **What are the drawbacks of actual loading?**

Ans. In that case machine has to run on load and the power consumed is merely a wastage. In case of huge motors the power loss will be high enough.

Q. **Why indirect method of testing is adopted?**

Ans. In huge machines, the energy required for direct testing is quite large and that is the wastage, to avoid that loss this test is preferred.

Q. **What is the Hopkinson's test?**

Ans. This test is also known as the Regenerative test. In this test two identical machines are taken. One machine is driven as a D.C. generator and other as a motor coupled with the same generator. This generated voltage is fed to the busbar. In this particular test, the beauty is that only the losses of both the machines are fed from the mains, rest of the energy is supplied by the same set.

Q. **What is the utility of this test?**

Ans. In this test, though we are feeding only the losses but actually both the machines are fully loaded under all normal conditions. Here the rise in temperature, winding resistances and other performance can be judged in a single test on full or at any desired load.

Q. **If both the machines are identical what formula is used for calculating the efficiency of the machine?**

Ans. $\eta = \sqrt{\dfrac{I_2}{I_2 + I_3}}$

where η = efficiency of the machine, I_2 = the current supplied by the generator and I_3 = current drawn from the mains.

TORQUE

Q. **Is there any relationship between the armature torque and armature current.**

Ans. Yes, $T_a \propto I_a$.

Q. **How shaft torque is calculated when B.H.P. and speed is given?**

Ans. $T_{sh} = \dfrac{\text{BHP} \times 735.5}{2\pi N}$.

Q. **Can you calculate the shaft torque if the iron and friction losses, back e.m.f. and speed are given?**

Ans. $T_{sh} = \dfrac{E_b \cdot I_a - \text{Iron and friction losses}}{2\pi N/60}$

Q. **What is the condition for maximum efficiency when losses are taken into consideration?**

Ans. When constant losses are equal to variable losses.

Q. **Can you calculate the load current for maximum efficiency?**

Ans. $I_l = \sqrt{\dfrac{P_a}{R_a}}$ for maximum efficiency.

Q. **What precautions must be taken for brake test for series motor?**

Ans. The brakes must be tight before the motor is started, otherwise the speed will be very high.

Q. **What do you understand from the retardation test.**

Ans. It is also known as running down test. It is used for finding out the stray losses of the shunt wound D.C. machines.

Q. **Name the I.S.I. classifications according to the rating of D.C. machines?**

Ans.
1. Continuous rating.
2. Continuous maximum rating.
3. Short time rating.

Q. **What is continuous rating?**

Ans. It is the output which a machine delivers continuously without exceeding the permissible temperature. It can deliver 25% overloading for 2 hrs.

Q. **Define continuous maximum rating.**

Ans. It is also similar to continuous rating except not allowing the overloading. It is used for motors of large capacity.

Q. **What is short time rating?**

Ans. It is the output which an electric machine can deliver for specified period without exceeding the specified temperature rise.

FAULTS

Q. **What will happen if in a motor the armature coil is opened?**

Ans. The motor will not start.

Q. **If the field circuit or coil is opened?**

Ans. The motor will run to a very high speed.

Q. **What can be the reasons, if the motor blows out the fuses at the time of starting?**

Ans.
1. Fuses of low capacity.

2. Overloading.
3. Earth fault.
4. Short circuit in starter resistance.
5. Short circuit in armature winding.

Q. **What are the reasons of heavy sparking on the commutator?**

Ans.
1. Dirty commutator.
2. Brushes are not properly seated.
3. Motor overloaded.
4. Brushes not in m.n.p. position.

Q. **If the motor starts at low speed what is the reason and how to remove it?**

Ans. It can be because of low voltage, defective bearing and wrong polarity of the adjacent poles. The supply voltage can be checked by means of the voltmeter, brushes should be replaced, if needed, and lubricate or grease the bearings and check the polarity of the poles and if wrong, correct it.

OBJECTIVE TYPE MULTIPLE CHOICE QUESTIONS

1. D.C. motor converts ...... into ......
   (a) mechanical energy; electrical energy
   (b) chemical energy; electrical energy
   (c) chemical energy; solar energy
   (d) D.C. electrical energy; mechanical energy.
2. The d.o.r. of the D.C. motor can be determined by ......
   (a) Fleming's right hand rule (b) Fleming's left hand rule
   (c) Ampere's rule (d) End rule.
3. The speed of D.C. shunt motor can be increased above normal by adopting the ...... method.
   (a) armature control (b) field control
   (c) voltage control (d) tapped on field control.
4. The speed of the D.C. motor is directly proportional to ......
   (a) armature current (b) field current
   (c) impressed voltage (d) number of poles
5. According to the field connections the D.C. motors are classified as ......
   (a) series, shunt and compound
   (b) series, universal and compound
   (c) shunt, compound and shaded pole
   (d) shaded pole, compound and series motor.

6. Three point starter is used to start the ......
   (a) D.C. shunt motor (b) series motor
   (c) universal motor (d) induction motor.
7. The speed of the D.C. motor on no load increases rapidly because of ......
   (a) open circuit in armature
   (b) open circuit in field winding
   (c) open circuit in line
   (d) open circuit in armature resistance.
8. The torque of D.C. motor is directly proportional to the ......
   (a) line voltage (b) line current
   (c) armature current (d) armature resistance
9. The resistance of N.V.R.C. is ......
   (a) high (b) low (c) medium (d) zero ohm.
10. In case of long shunt compound motor the series field winding is connected ......
    (a) in series with shunt field (b) across the armature
    (c) in series of only armature (d) None of the above.
11. To minimise the armature reaction in motor, the brushes are ......
    (a) kept on the same segment as before
    (b) given forward lead in the direction of rotation of the motor
    (c) given backward lead to the d.o.r. of motor
    (d) not disturbed.
12. The sparking on the commutator of a normal D.C. motor is because of the ......
    (a) current in field winding
    (b) current in armature and field winding
    (c) reactance voltage
    (d) number of poles.
13. The resistance of O.L.R.C. is ...... and is connected in ...... with the line.
    (a) high; parallel (b) low, parallel
    (c) low, series (d) high series.
14. The back e.m.f. of a D.C. motor depends upon ......
    (a) field flux and poles
    (b) speed
    (c) number of armature conductors
    (d) All of these.
15. In the absence of back e.m.f. the motor will take ......
    (a) no current (b) heavy current
    (c) low current (d) full load current.

16. State true or false.
    (a) The starting resistance of a D.C. motor is usually a low resistance.
    (b) In starting motor does not have any back e.m.f.
    (c) The speed of D.C. shunt motor falls very rapidly with the increase of load.
    (d) The series motor has approximately a constant speed.
    (e) D.C. series motor has a centrifugal switch.
    (f) Hopkinson's test can be performed only on A.C. series motor.
    (g) The eddy current losses can be less if the laminations are thinner.
    (h) Pole shoes of D.C. motor are never laminated.
    (i) The back e.m.f. is $E_b = V - I_a \cdot R_a$.
    (j) Traction load requires high starting torque.
17. In case of D.C. three point starter the overload tripping contact actually short circuits the ......
    (a) overload release coil (b) no volt release coil
    (c) starting resistance (d) field winding.
18. To bring back to 'OFF' position a ...... is provided with the handle.
    (a) spring (b) lever (c) iron strip (d) copper strip
19. In a four point D.C. motor starter the no volt release coil circuit is ......
    (a) in series with the armature (b) in series with the shunt field
    (c) independent (d) short-circuited.
20. To limit the short circuit current while motor is protected against overloading a ...... is provided in the no volt coil independent circuit.
    (a) inductor (b) resistor
    (c) capacitor (d) tripping contact.

CHAPTER 15

Alternating Current

INTRODUCTION

Q. **Define alternating quantities.**

Ans. The quantity whose magnitude undergoes a definite cycle of change during definite interval of time, is known as alternating quantity.

FUNDAMENTAL TERMS AND DEFINITIONS

Q. **What do you understand by the waveform?**

Ans. The curve obtained by plotting the instantaneous values of alternating quantity as ordinate against time as abscissa is known as waveform.

Q. **What is one cycle?**

Ans. The complete change in value and direction of an alternating quantity is called cycle.

Q. **How many electrical degrees are there in one cycle?**

Ans. 360° elect.

Q. **What do you mean by electrical degrees?**

Ans. One pair of poles is supposed to have 360° elect and total electrical degrees are equal to

° elect = mechanical degrees × pair of poles

= 360 × pair of poles

Q. **How many maximum electrical and mechanical degrees are possible?**

Ans. Mechanical degrees are maximum 360° but electrical degrees can be the multiple of 360° depending upon the number of poles of the machine.

Q. **What is sine wave?**

Ans. Whenever the instantaneous values of alternating quantity are plotted against the sine of the angle covered, the curve so obtained is known as sine wave.

Q. **What is frequency?**

Ans. The number of cycles per second is known as the frequency. It is denoted by letter '*f*' or Hz and unit is cycles/sec.

Q. **Define the time period?**

Ans. It is the time taken to complete one cycle. It is denoted by *T*.

Q. **Is there any relationship between the frequency and time period?**

Ans. Yes, $T = \frac{1}{f}$.

Q. **What is the frequency of supply in India?**

Ans. 50 c/s.

Q. **Is there any country which has 60 c/s frequency?**

Ans. Yes, in U.S.A.

Q. **What is the maximum value or peak value or amplitude of the alternating quantity?**

Ans. The maximum value attained on either positive or negative side of an alternating quantity is called the maximum value or amplitude or peak value of alternating quantity.

Q. **What is the instantaneous value of alternating quantity?**

Ans. The value of alternating quantity at any instant during one cycle is known as the instantaneous value at that particular instant.

Q. **What is the average value of alternating quantity?**

Ans. It is the average of all the instantaneous values attained during half cycle either positive or negative.

Q. **How it is related to the maximum value?**

Ans. $I_{av} = 0.637 I_{\max}$

or it is equal to the 0.637 of the maximum value of alternating quantity.

Q. **Is there any practical utility of the average value of A.C. in current electricity?**

Ans. No.

Q. **What is the r.m.s. value of alternating quantity?**

Ans. The r.m.s. value of alternating quantity is defined as the quantity which when applied to a given circuit for a given time produces the same expenditure of energy as when D.C. supplied to the same circuit for the same interval of time.

It is given as $I_{rms} = 0.707 I_{\max}$.

Q. **What is the utility of r.m.s. value of A.C.?**

Ans. All the machines, instruments and other electrical appliances are rated in r.m.s. value of alternating current and even the meter reads only the r.m.s. value.

Q. **Define form factor?**

Ans. It is the ratio of r.m.s. value to average value of alternating quantity.

$$k_f = \frac{\text{r.m.s. value of alternating quantity}}{\text{average value of alternating quantity}} = 1.11$$

Q. **Define peak factor.**

Ans. It is the ratio of maximum value and r.m.s. value of alternating quantity

$$\text{Peak factor} = \frac{\text{Max. value of alternating quantity}}{\text{r.m.s. value of alternating quantity}}$$
$$= 1.414$$

Q. **What is the phase difference?**

Ans. The relative time of change in value of two alternating quantities is called phase difference.

OR

The angular displacement between two alternating quantities is called the phase difference.

Q. **If two alternating quantities attain their maximum and minimum values at the same time, what relationship these possess?**

Ans. In phase.

Q. **What is meant by the out of phase?**

Ans. If two alternating quantities do not attain their maximum and minimum values at the same time these are termed to be out of phase.

Q. **What is impedance?**

Ans. The total opposition in the flow of alternating current is known as impedance. The impedance is generally denoted by letter Z and measured in ohms.

Q. **What is the reciprocal of the impedance?**

Ans. Admittance.

Q. **What is inductive reactance?**

Ans. It is the opposition offered because of the inductance in the flow of electric current. It is denoted by letter X_L and measured in ohms.

Q. **What is the relationship between the inductance and inductive reactance?**

Ans. $X_L = 2\pi fL$ ohms.

Q. **What do you mean by the susceptance?**

Ans. The susceptance is the reciprocal of the reactance and is denoted by letter s or b.

Q. **How the capacitive reactance is denoted and defined?**

Ans. It is the opposition because of the capacitance or capacitive reactance. It is denoted by letter X_C.

Q. **How X_C is calculated?**

Ans. $$X_C = \frac{1}{2\pi fC}\ \Omega.$$

Q. **Obviously the total opposition is known as the impedance, what are the main factors which actually constitutes the total opposition?**

Ans. Resistance, inductance and capacitance.

Q. **What is power factor?**

Ans. It is the ratio of true power to apparent power. The angular displacement between two alternating quantities is known as the phase displacement and cosine of this angle of displacement is known as power factor.

OR

In series circuit the ratio of resistance and impedance is called power factor.

Q. **What do you mean by lagging and leading power factor?**

Ans. If the current is lagging behind the voltage, then power factor attained is called lagging power factor.

If the current leads the voltage then the power factor is leading power factor.

VECTOR

Q. **What are the vector quantities?**

Ans. The quantity which has got the magnitude and direction both, is known as the vector quantity.

Q. **What do you understand by the scalar quantity?**

Ans. The quantity which has either the magnitude or direction is known as scalar quantity.

Q. **Are the A.C. quantities like current and voltage, vector quantities?**

Ans. Yes, these are the vector quantities and can be represented by vectors.

Q. **If there are two alternating quantities (a) at right angle to each other, (ii) having 180° displacement, (iii) having zero degree displacement, and (iv) having any other angular displacement can you find out their resultant?**

Ans. (i) If the quantities are at right angle to each other then

$$R = \sqrt{A^2 + B^2}$$

(ii) If these are 180° apart

$R = A - B$ (algebraic)

(iii) If these are at 0°

$R = A + B$ (algebraic)

(iv) If any angle say θ°

$$R = \sqrt{A^2 + B^2 + 2AB\cos\theta}$$

Or these can be resolved into vertical and horizontal components and their resultant can be determined as

$R^2 = (\Sigma H)^2 + (\Sigma V)^2$

Q. **Can you tell the angle by which both current and voltage differ?**

Ans. $\tan \phi = \dfrac{\Sigma V}{\Sigma H}$

thus the angle ϕ is known from the log tables.

Q. **Where the admittance method is suitably or conveniently used?**

Ans. In parallel circuits for calculating total current and power factor of the load.

Q. **Can you tell the Ohm's law in A.C. circuit?**

Ans. It is, Current = Voltage/Impedance.

Q. **Does the resistance of a conductor depend upon the frequency?**

Ans. No, it is independent of the frequency.

A.C. CIRCUITS

Purely Resistive Circuits

Q. **What is the current and voltage relationship in the purely resistive circuit?**

Ans. Both are in phase.

Q. **Will the resistive load consume power?**

Ans. Yes.

Q. **What is the formula for calculating the power in case of resistive load?**

Ans. Power = Voltage × Current.

Q. **Why power factor in the resistive circuit is unity?**

Ans. The power factor is the cosine of the angle between current and voltage. Here both are in phase having no displacement hence cos 0 is one and power factor is unity.

Purely Inductive Circuit

Q. **What is purely inductive circuit?**

Ans. The inductive circuit in which neither the resistive nor capacitive nature of load is connected, is known as purely inductive circuit.

Q. **What is the relationship between the current and voltage in the purely inductive load?**

Ans. Out of phase. Current is lagging behind the voltage by 90°.

Q. **What is the value of power factor in the purely inductive circuit?**

Ans. It is zero.

Q. **Does an inductive circuit draw any power from the supply?**

Ans. No.

Q. **While drawing the sinusoidal waveform of current and voltage, the power line also comes into existence, what does it indicate?**

Ans. It indicates that though power is shown, but during one complete cycle, the power is consumed twice and delivered to the same circuit twice thus total power consumed during one cycle is zero.

Q. **Define inductance.**

Ans. The inductance is the property of the circuit by virtue of which an e.m.f. is induced in the same circuit, it opposes the cause responsible for the production of this e.m.f.

Q. **What is the unit of inductance?**

Ans. Henry.

Q. **If the frequency of supply is doubled what will happen to the inductive reactance?**

Ans. It will be doubled.

Q. **There are two coils, one of having 0.01 H and other 0.001 H inductance, respectively. If these are connected across the same voltage, which coil will take more current?**

Ans. The coil having inductance of 0.001 H.

Purely Capacitive Circuit

Q. **What is the purely capacitive circuit?**

Ans. The circuit which has neither the resistive nor inductive components but only capacitance, is known as purely capacitive circuit.

Q. **What is the relationship between current and voltage in purely capacitive circuit?**

Ans. These are out of phase. Current is leading the voltage by 90°.

Q. **Can you tell, what will be the reading of the wattmeter connected to purely capacitive circuit?**

Ans. Zero.

Q. **How will you calculate the capacitive reactance?**

Ans. $X_C = \frac{1}{2\pi fC}$ ohms.

Q. **In which units the capacity of a condenser is measured?**

Ans. Farad or microfarad.

Q. **What is one microfarad, picofarad?**

Ans. 10^{-6} farad = one microfarad

10^{-12} farad = one picofarad.

A.C. SERIES CIRCUIT

Q. **Does the power factor of a series circuit always remain lagging?**

Ans. No, it depends upon the components connected and their values.

Q. **Why power factor is lagging in series circuit?**

Ans. It is because the circuit has reactive components. If both capacitive and inductive loads are there, and the inductive component is more than the capacitive component in the circuit.

R.L. Series Circuit

Q. **What is R.L. series circuit?**

Ans. The circuit in which resistance and inductance are connected in the form of a chain across the voltage, forming only one path of the current to flow, is known as the R.L. series circuit.

Q. **Will the current in R and L elements be same?**

Ans. Yes, there is only one path of current, the current will be V/Z amperes.

Q. **How will you calculate the voltage drop across each element?**

Ans. Voltage across resistance = Current × Resistance

Voltage across inductance = Current × Reactive inductance.

Q. **How much is the angular displacement between the voltage drop across resistance and inductance?**

Ans. 90°.

Q. **How will you calculate the voltage fed to the circuit if voltage drops V_R and V_L are given?**

Ans. $V = \sqrt{V_R^2 + V_L^2}$

Q. **What will be the power factor in A.C. series circuit?**

Ans. The ratio of resistance to impedance.

i.e. $\cos \phi = \dfrac{R}{Z}$

Q. **What will be the impedance in R.L. series circuit?**

Ans. $Z = \sqrt{R^2 + X_L^2}$ ohms.

Q. **What is the power consumed?**

Ans. Voltage × Current × Power factor

R.C. Series Circuit

Q. **What is R.C. series circuit?**

Ans. If the purely resistive and capacitive elements are so connected as to form a chain, these are said to be connected in series and the circuit is called R.C. series circuit.

Q. **What is the current and voltage relationship in R.C. series?**

Ans. Current is leading the voltage.

Q. **What is the value of current?**

Ans. $\text{Current} = \dfrac{\text{Voltage}}{\text{Impedance}}\ \text{amp} = \dfrac{V}{Z}\ \text{amp.}$

Q. **What is the total impedance in R.C. series circuit?**

Ans. $Z = \sqrt{R^2 + X_C^2}$.

R.L.C. Series Circuit

Q. **What do you know about the general series circuit?**

Ans. The circuit which has all the three—resistive, inductive and capacitive—elements connected in series, is known as general series circuit.

Q. **What is the voltage drop in resistive, inductive and capacitive elements?**

Ans. $V_R = I \times R$, $V_L = I \times X_L$ and $V_C = I \times X_C$

Q. **Can you tell the angular displacement between V_C and V_L?**

Ans. These are perfectly 180° apart.

Q. **Can you increase the power factor of R.L.C. series circuit?**

Ans. Yes, by increasing the capacitive element in the circuit. Or by decreasing the inductive element from the circuit.

Q. **While you are asked to draw the vector diagram of R.L.C. series circuit, generally which is taken as reference vector?**

Ans. Current, because it is common in all components.

RESONANT FREQUENCY AND POWER FACTOR

Q. **What do you understand by the resonant frequency?**

Ans. The frequency on which the X_L and X_C become same is known as the resonant frequency.

Q. **A general series circuit is taking maximum current, can you tell about the frequency of the circuit?**

Ans. The circuit is operating at resonant frequency.

Q. **What is the formula for calculating the resonant frequency?**

Ans. $f_r = \dfrac{1}{2\pi\sqrt{LC}}$

Q. **How frequency is related with the X_L and X_C.**

Ans. $X_L \propto f$ and $X_C \propto \dfrac{1}{f}$

Q. **What do you mean by the true power?**

Ans. It is the actual power which is used for doing useful work. It is equal to the product of current, voltage and power factor.

Q. **What is apparent power?**

Ans. It is power which appears to be consumed and is equal to the product of current and voltage.

Q. **Is there any other kind of power except true and apparent.**

Ans. Yes, it is reactive power.

Q. **What is reactive power?**

Ans. It is the power which is used in the reactive components of the circuit or say this is the product of current, voltage and sine of the angle of lead or lag.

Q. **What is the relationship between the true power, apparent power and reactive power?**

Ans. Apparent power $= \sqrt{(\text{True power})^2 + (\text{Reactivepower})^2}$

Q. **If you are required to improve the power factor of the load, which component (either the reactive or wattful) is to be reduced?**

Ans. Reactive component.

Q. **What do you mean by the impedance triangle?**

Ans. The triangle whose sides represents resistance, reactance and impedance of a circuit, is called the impedance triangle.

Q. **Generally which side of a impedance triangle shows the impedance?**

Ans. Hypotenuse.

Q. **If the sides of a right angle triangle shows the useful power, and apparent power, what do you call this triangle?**

Ans. Power triangle.

R.L. PARALLEL CIRCUIT

Q. **What is R.L. parallel circuit?**

Ans. If the resistance and inductance are connected in parallel, the circuit is called the R.L. parallel circuit.

Q. **What is common in parallel circuit—the current or voltage?**

Ans. Voltage.

Q. **How the total current is calculated?**

Ans. If I_R is the current in resistance, I_L the current in inductance, then total current $I = \sqrt{I_R^2 + I_L^2}$.

Q. **What is the total impedance of the circuit?**

Ans. $\frac{1}{Z^2} = \frac{1}{R^2} + \frac{1}{X_L^2}$.

Q. **What is power factor in this circuit?**

Ans. $\frac{Z}{R}$.

R.L.C. PARALLEL CIRCUIT

Q. **How the current and voltage behaves in :**
(a) resistance
(b) inductance and
(c) capacitance?

Ans. (a) In phase with the voltage.
(b) Out of phase, current lagging behind the voltage.
(c) Out of phase, current leading the voltage.

Q. **What is the relationship between V_R and V_L?**

Ans. These are out of phase by 90°.

Q. **What is the formula for calculating the impedance in general series circuit?**

Ans. $\frac{1}{Z^2} = \frac{1}{R^2} + \left(\frac{1}{X_L} \sim \frac{1}{X_C}\right)^2$.

Q. **What is power consumed in the circuit?**

Ans. $W = V \times I \times \cos \phi$.

Q. **In a case of parallel resonance what will be the current drawn?**

Ans. The minimum possible.

Q. **In series resonance what is the value of current?**

Ans. The maximum possible.

Q. **What is the formula for calculating the inductance in resonant circuit?**

Ans.
$$L = \frac{1}{4 \cdot \pi^2 \cdot C \cdot f_r^2}.$$

Q. **What is the dynamic impedance of the parallel circuit?**

Ans. It is $\frac{L}{CR}$.

Q. **What is the power factor in case of general parallel circuit?**

Ans. The ratio of impedance to equivalent resistance of the circuit.

or $\tan \phi = \frac{\Sigma V}{\Sigma H}$ of all currents :

See for ϕ from log table and find $\cos \phi$.

Q. **For the same power if you increase the power factor, what effect does it exhibit on the current?**

Ans. Current drawn for the same power will decrease.

Q. **For the same power if the power factor is further decreased what effect does it make to the current drawn?**

Ans. The alternator, conductor, motor and other appliance, where power factor is also an effective factor, will draw more current to meet the same power.

Q. **Does the power factor of the system depend upon the load connected or the generated voltage?**

Ans. Load connected.

Q. **How the power factor is improved?**

Ans. (a) By connecting the bank of capacitors.

(b) By using a synchronous capacitor i.e. overexcited synchronous motor.

Q. **What do you mean by the bank of capacitors?**

Ans. The number of capacitors are connected in parallel. As there are more number of capacitors connected, so it is known as bank of capacitors.

Q. **Actually speaking, what these contribute to improve the power factor?**

Ans. It minimise the reactive component of the power.

Q. **In case when the overexcited synchronous motor is run to improve the power factor, what we call it?**

Ans. Synchronous capacitor.

POLYPHASE SYSTEM

Q. **What is meant by the polyphase supply system?**

Ans. Supply having more than one phase is known as polyphase supply system.

Q. **How many phases supply is generally produced?**

Ans. Three phase.

Q. **What is the angular displacement between the phases in case of 3-ϕ supply?**

Ans. 120°.

Q. **Can it be possible to generate six phase supply?**

Ans. Yes.

Q. **How many windings are there in a 3-ϕ alternator?**

Ans. Three sets of winding.

Q. **What are the possible methods of inter-connecting the three windings for three phase supply?**

Ans. These are three phase six wires, three phase three wires and three phase four wires systems.

Q. **How many windings are necessary for producing two phase supply?**

Ans. Two windings.

Q. **What will be the line voltage in case of two phase supply if the phase voltage is 100 V?**

Ans. 141.4 volts.

Q. **What will be the power in case of 2-ϕ system?**

Ans. $W = \sqrt{2} V_L \cdot I_L \cdot \cos\phi$.

Q. **What do you mean by star point?**

Ans. If three similar ends of the phase windings are connected to a common point, the point is known as star point or neutral point.

Q. **What is the relationship between current and voltage in star connections?**

Ans. Line current = Phase current

Line voltage = $\sqrt{3}$ × Phase voltage.

Q. **Where the phase voltage is measured?**

Ans. Between the line and neutral or star point.

Q. **Can you tell the formula for finding out the power connected in three phase?**

Ans. $W = \sqrt{3} V_L \cdot I_L \cdot \cos\phi$.

Q. **What is three phase three wire system?**

Ans. The system in which neutral is not available and only three lines are drawn. These are known as delta connections.

Q. **Which ends are connected in case of delta connection?**

Ans. Dissimilar ends.

Q. **What are the relationships between current and voltage in case of delta connections?**

Ans. Line voltage = Phase voltage.

Line current = $\sqrt{3}$ × Phase current.

Q. **How will you calculate the power in case of delta connection?**

Ans. $W = \sqrt{3} V_L \cdot I_L \cdot \cos\phi$.

Q. **Can the load be operating in case of open delta?**

Ans. Yes, with a 60% efficiency.

Ans. In case of delta connections, if one phase winding is broken or open circuited the connections are known as open delta connection.

Q. **What are the balanced and unbalanced loads?**

Ans. If the load on all three phases is same, i.e. the load current flowing through each phase is same that is known as balanced load if it is not same then it is known as the unbalanced load.

Q. **How the power in case of three phase system is measured?**

Ans.
1. By three wattmeters of single phase.
2. By two wattmeters.
3. By single wattmeter of single phase (balanced load).
4. By single wattmeter of three phase.

Q. **How the wattmeters are connected?**

Ans. The current coils in series with the load and the pressure coil across the supply voltage. It can be across two phases or one phase and neutral.

Q. **If three single phase wattmeters are connected to measure the actual wattage consumed what will be the power consumed. The readings are W_1, W_2 and W_3, respectively.**

Ans. $W_{Total} = W_1 + W_2 + W_3$ watts.

Q. **In case of single wattmeter of single phase, how will you connect the pressure coil in case of delta system?**

Ans. An artificial neutral will be prepared and the pressure coil will be connected across the one phase and neutral.

Q. **How artificial neutral is prepared?**

Ans. By connecting two more resistances having the same value as that of pressure coil and thus forming an artificial star.

Q. **What are the methods of power measurement in case of three phase unbalanced loads?**

Ans.
1. Two wattmeter method.
2. One wattmeter of three phase.
3. Three wattmeters of single phase.

Q. **In case of two wattmeter method, the pressure coil of these wattmeters is designed for how much voltage?**

Ans. 440 V.

Q. **What will be the total power in case of two wattmeters method?**

Ans. $W_T = W_1 + W_2$.

Q. **Can the p.f. of the load be calculated by two wattmeter method?**

Ans. Yes, first calculate tan ϕ, see for ϕ from log tables and note the value of cos ϕ.

Here $\tan \phi = \sqrt{3}\frac{W_1 - W_2}{W_1 + W_2}$.

Q. **If the load power factor is 0.5, will the reading be obtained in both wattmeters?**

Ans. No, only one wattmeter will give reading the second will give zero reading.

Q. **If the power factor is below 0.5, what will be the position of both wattmeters?**

Ans. One of them will give opposite (reverse) reading.

Q. **How to get positive reading?**

Ans. Reverse the connection of p.c. or c.c. of that wattmeter.

Q. **In which condition both wattmeters will give positive reading?**

Ans. When the power factor is above 0.5.

OBJECTIVE TYPE MULTIPLE CHOICE QUESTIONS

1. A pure resistive circuit in A.C. follows the ......
   (a) Ohm's law (b) Faraday's laws of electrolysis
   (c) Lenz's law (d) Kirchhoff's law.
2. An alternating current changes its ...... and ...... in a definite interval of time.
   (a) value; direction (b) magnitude; phase
   (c) angular displacement; direction
   (d) instantaneous value; power.
3. The frequency of A.C. in India is ......
   (a) 50 c/s (b) 60 c/s (c) 40 c/s (d) 25 c/s
4. The time period is ...... to the frequency of A.C. supply.
   (a) directly proportional (b) inversely proportional
   (c) sixty times (d) fifty times.
5. The ...... value of A.C. is useful for all practical purposes.
   (a) maximum (b) effective
   (c) instantaneous (d) average.
6. The average and effective values of A.C. are ...... respectively.
   (a) 0.707 and 0.637 to maximum value
   (b) 0.637 and 0.707 to maximum value
   (c) 1.414 and 0.707 to maximum value
   (d) 0.307 and 0.637 to maximum value.
7. The current and voltage in purely resistive load are ......

(a) having 90° elect. angular displacement
(b) having zero degree phase displacement
(c) having 40° elect. phase displacement
(d) having 270° elect. phase displacement.

8. In purely inductive load on A.C. the current ...... the voltage by ...... electrical.
(a) leads; 0° (b) lags; 0° (c) leads; 90° (d) lags; 90°

9. Current in purely capacitive circuit ...... the voltage by ...... degree electrical.
(a) leads; 0° (b) lags; 0° (c) leads; 90° (d) lags; 90°

Polyphase System

10. The angular displacement between the phases of a polyphase system is ......
(a) 120° (b) 0°
(c) 360° divided by the number of phases except two phase system
(d) 90°

11. In most of the cases the polyphase supply is generated as ......
(a) single phase supply (b) two phase supply
(c) three phase supply (d) six phase supply.

12. Most common three phase inter-connections are ......
(a) series and parallel (b) star and delta
(c) series and star (d) delta and open delta.

13. In star connections the phase current is ...... the line current.
(a) $\sqrt{3}$ times (b) 0.707 times
(c) same (d) equal and opposite.

14. In case of star connections the line voltage is equal to ...... times the phase voltage.
(a) 0.707 times (b) 0.637 times
(c) 1.732 times (d) 1.414 times.

15. In delta connections the line voltage is ...... the phase voltage.
(a) 1.414 times (b) 0.707 times
(c) $\sqrt{3}$ times (d) equal

16. Power in three phase balanced load can be given as ......
(a) $\sqrt{3}$ × line voltage × line current × power factor
(b) $\sqrt{3}$ × phase current × phase voltage × power factor
(c) 3 × line voltage × line current × power factor
(d) 3 × phase voltage × line voltage.

17. In three phase balanced load, the power can be measured with minim ...... number of wattmeters.

(a) one (b) two (c) three (d) four.

18. The two wattmeter method is used to measure three phase ...... load.
(a) balanced load
(b) unbalanced load
(c) balanced and unbalanced loads
(d) only resistive load.

19. In case of two wattmeter method of power measurement, if one wattmeter reads zero the load power factor is ......
(a) 0.5
(b) less than 0.5
(c) more than 0.5
(d) unity.

20. The negative reading of one wattmeter in case of two wattmeter method of power measurement indicates the power factor ......
(a) 0.5
(b) unity
(c) less than 0.5
(d) more than 0.5.

Power

21. The electrical power can be calculated by using the formula ......
(a) $P = IR \cos \phi$
(b) $P = VI \cos \phi$
(c) $P = IR^2 \cos \phi$
(d) $P = IR \cos \phi$

22. The electrical power is expressed as ......
(a) $P = I^2R$
(b) $P = IR^2$
(c) $P = \sqrt{IR}$
(d) $P = V^2R$

23. In case voltage and resistance are given the power is ......
(a) $P = \dfrac{V^2}{R}$
(b) $P = \dfrac{V}{R}$
(c) $P = \dfrac{V}{R^2}$
(d) $P = \sqrt{\dfrac{V}{R}}$

24. Electrical power consumed can be calculated by the following three formulae ......
(a) $P = VI, P = \dfrac{V^2}{R}$ and $P = I^2R$
(b) $P = VI, P = \dfrac{I^2}{V}$ and $P = V^2 \times I$
(c) $P = IR, P = \dfrac{V^2}{R}$ and $P = VI$

25. Two resistances 3 Ω and 7 Ω are connected in series across 30 volts the power consumed is ......
(a) 240 W (b) 120 W (c) 90 W (d) 72 W

26. To convert mechanical power into electrical power, multiply the power by ......
    (a) 746 (b) 846 (c) 946 (d) 646.
27. Rheostats are usually rated in ......
    (a) watts and ampere (b) watts and volt
    (c) resistance and ampere (d) ampere and watts.
28. Two lamps of 40 watts and 60 watts are connected in parallel their resultant power is ......
    (a) 100 watts (b) 20 watts (c) 50 watts (d) 240 watts
29. If a D.C. 240 V is connected across 240 Ω resistance the power of the load is ......
    (a) 240 watts (b) 1 watt (c) 10 watts (d) 480 watts
30. If 6 resistances of 60 Ω each are connected in parallel the total resistance will be ......
    (a) 10 Ω (b) 30 Ω (c) 360 Ω (d) 90 Ω

Power Factoı

31. The power factor is defined as the ...... of the angle between voltage and current.
    (a) cosine (b) sine (c) tangent (d) sec.
32. In case of series circuit the power factor is given by the ratio of ......
    (a) resistance to inductance (b) resistance to capacitance
    (c) resistance to impedance (d) capacitance to inductance
33. The power factor in any circuit is defined as the ratio of ......
    (a) true power to apparent power
    (b) true power to inductive power
    (c) apparent power to inductive power
    (d) single phase to three phase power.
34. In parallel circuit the power factor is the ratio of ......
    (a) capacitance to inductance
    (b) resistance to impedance
    (c) equivalent impedance to equivalent resistance
    (d) reactance to resistance
35. The power factor of A.C. purely resistive circuit is ......
    (a) unity (b) 90° lagging
    (c) 90° leading (d) 60° leading.
36. In A.C. circuit when the power factor is unity the reactive component of the circuit is ......
    (b) zero

(c) equal to resistive component
(d) equal to r.m.s. component of the sine wave.

37. The power factor of a circuit can be improved by ......
(a) increasing the reactive component of the circuit
(b) decreasing resistive component of the circuit
(c) increasing the real power supplied
(d) decreasing the reactive component.

38. The power factor improvement may be achieved by using ......
(a) long transmission line (b) short transmission line
(c) the induction motor (d) synchronous motor.

39. A consuming unit has a load of 1000 kW and the reactive power 1000 kV. A.R., it has the power factor ......
(a) unity (b) 0.8 (c) 0.9 (d) 0.7

40. The power factor of a circuit can be obtained by using ......
(a) ammeter (b) voltmeter
(c) power factor meter (d) phase sequence indicator.

41. Insert the correct word to complete the sentence.
(i) The voltage and current in resistor are ...... and in inductance are ......
(ii) In pure inductive circuit the current lags the voltage drop across the coil by ......
(iii) the resistance and reactance must be added ...... to obtain the value of impedance.
(iv) The only component in RLC circuit that takes power is the ......
(v) The circuit in which $X_L = X_C$ in RLC series circuit is known as ......

42. Select the best suited answer to complete each statement.
(i) Frequency is measured in ......
(ii) The power factor is measured in ...... unit because it is a ......
(iii) In case of resonance $X_C =$ ......
(iv) Power is measured in ......
(v) The power in purely inductive circuit is measured by ......
(vi) At unity power factor a ...... amount of current is required to deliver a given amount of power.
(vii) The inductive reactance of an air-cored coil may be increased by inserting an ......
(viii) If in a R.L.C. series circuit the current is maximum, the circuit is ......
(ix) The apparent power can be measured by ......

CHAPTER 16

Alternator

INTRODUCTION

Q. **What is an alternator?**

Ans. The machine which converts the mechanical energy into A.C. electrical energy is known as alternator.

Q. **What are the main parts of an alternator?**

Ans. Stator, rotor, fan, slipring, side covers etc.

Q. **What are the three essential elements for generating the e.m.f.?**

Ans. The source of prime mover or say motion, magnetic field and armature conductors.

Q. **What are the sources of prime mover generally used for the alternators?**

Ans. The water turbine, steam turbine and diesel engine.

Q. **How the magnetic field is produced?**

Ans. The magnetic field is produced by the magnetic poles. The field can be of two types—the stationary and rotating.

Q. **Generally which type of field is used for the alternators?**

Ans. Rotating field type.

Q. **What are the main rotating parts of an alternator?**

Ans. Armature, sliprings, fan, bearings, pulley, excitor etc.

Q. **Name the stationary parts of the alternator?**

Ans. The stator winding (may be poles in case of stationary pole-type construction or three phase armature winding for rotating pole type), side covers, rocker and fingers, body or yoke, bed sheet etc.

Q. **Which type of alternators are common in use?**

Ans. Rotating field type.

Q. **Why stationary field type alternators are not preferred because of design aspects?**

Ans. The space of three phase armature winding is more in stationary armature type construction, because of more circumferential area, more space for providing insulation, and winding, more space also facilitate for low dielectric strength insulation suitability, more conductors can be easily housed (increasing generated voltage) working space is more etc.

Q. **Is there any specific advantage about load current and voltage induced?**

Ans. Yes, when the armature is fixed the load current will not create any sparking as in case of rotating armature type construction. The sliding contact is dangerous than a fixed one.

Q. **What is the function of an excitor?**

Ans. An excitor is used to excite the field winding of the alternator for producing magnetic flux.

Q. **Will an excitor induce only a fixed voltage?**

Ans. No, the voltage induced can be changed by adding/or removing a field resistance in the field winding.

Q. **Which way adding or removing of external resistance in the field winding helps the alternator?**

Ans. If the load on the alternator is more, the voltage drop will result. Hence to increase the voltage or compensate the voltage drop, the field current has to be increased causing increase in induced voltage i.e. emf simultaneously for lowering the voltage we have to reduced the field current (when the load current is less and voltage induced is more).

Q. **Which method of voltage control is used?**

Ans. Field control method.

Q. **Can you name the general type of D.C. generator used as an excitor?**

Ans. It is D.C. shunt generator.

Q. **The sliprings are made of which metal?**

Ans. Gunmetal.

Q. **How will you collect the emf (rotary armature type) or feed the current (rotary field type) from the rotor circuit?**

Ans. Through collecting fingers (rotary armature) and the brushes.

Q. **What is the utility of the fan?**

Ans. The fan is provided for air circulation, the heat produced is dissipated and the alternator is kept cool.

Q. **Can you give some idea about the instruments mounted on the panel board of a three phase alternator?**

Ans. The panel board is supposed to have three ammeters MI type for A.C.

measurement, three voltmeters (MI type) for line voltage measurement (a rotating switch may also be provided to note the line voltage in that case one switch and one voltmeter or ammeter is sufficient), a frequency meter, a power factor meter, a wattmeter, an energymeter, a D.C. voltmeter, a D.C. ammeter, two circuit breakers, two field rheostats for alternator and excitor.

Q. **What is synchroscope?**

Ans. It is an instrument used for synchronizing the alternators.

Q. **What is the basic principle behind this instrument?**

Ans. Relative speed of two rotating magnetic fields.

Q. **How these fields are obtained?**

Ans. One because of the already running supply and other by the e.m.f. induced of the incoming machine.

Q. **Which load is known for its magnetising characteristics?**

Ans. Leading power factor load current has magnetic characteristic.

Q. **Lagging power factor load is known for its behaviour as ......?**

Ans. Demagnetising nature.

WORKING PRINCIPLE

Q. **What is the working principle of the alternator?**

Ans. The alternator works on the principle of Faraday's laws of electro-magnetic induction.

"Whenever a conductor cuts the line of force (magnetic flux) an e.m.f. is induced in that conductor."

Q. **Where the conductors, in which e.m.f. is induced, are placed?**

Ans. Generally in the stator.

Q. **How the connections are made, in star or in delta?**

Ans. Generally in star.

Q. **If a generator has four sliprings, name the type of generator?**

Ans. Stationary field type alternator or rotating armature type alternator.

TYPES

Q. **How the alternators are classified?**

Ans. These are classified in accordance with :

(i) the field construction
(ii) the number of poles
(iii) the types of field excitation
(iv) the types of pole construction.

Q. **Name the types of alternators according to the field.**

Ans. 1. Stationary field type.
2. Rotating field type.

Q. **What is the difference between the rotary and stationary field type construction?**

Ans. The alternators in which the magnetic field is stationary are called stationary field and those in which field in rotating are called rotating field type.

Q. **What are the advantages of rotating field type alternator?**

Ans. 1. Only two sliprings are required.
2. Winding can be properly insulated and housed.
3. More number of armature conductors can be placed as more space is available in stator than rotor.
4. Winding (armature) is stationary so less possibility of faults.
5. Less maintenance cost.

Q. **According to the pole construction, what are the types of alternators?**

Ans. 1. Salient pole type.
2. Smooth cylindrical type.

Q. **What do you mean by salient pole type?**

Ans. The poles are projected on the rotor. These types of rotor have more diameter than the axial length.

Q. **Can this type of rotor be run at fast speed?**

Ans. No.

Q. **What are the smooth cylindrical type poles?**

Ans. In this type the alternator has less diameter and more axial length. In this construction the projection of poles is totally avoided and rotor is of smooth cylindrical nature.

Q. **How many poles are generally designed in smooth rotor construction?**

Ans. Two or four.

Q. **Is this construction suitable for high speed?**

Ans. Yes.

Q. **Can you tell the number of poles designed for salient pole type rotor construction?**

Ans. Six pole to 48 poles.

Q. **How the alternators are classified according to the number of phases?**

Ans. Single phase, three phase etc.

Q. **How the field winding is excited in an alternator?**

Ans. 1. By means of a separate battery.

2. By means of a D.C. generator mounted on the same shaft.

Q. **Which type is preferred?**

Ans. Generator mounted on the same shaft.

Q. **What is the main advantage of this excitor?**

Ans. Economy of space.

Q. **Is there any disadvantage also?**

Ans. Yes, the excitor has to run on same speed as that of alternator.

Q. **In general the speed of these D.C. shunt generators is not so fast so what about the mechanical size of this excitor and the normal D.C. generator?**

Ans. In comparison these are made smaller in size.

Q. **How the poles are fixed in the rotor of an alternator having speed less than 25 m/sec?**

Ans. Bolted with rotor.

Q. **If the speed is higher than 25 m/sec how these are fixed?**

Ans. With the dovetailed joints with the rotor.

E.M.F. EQUATION

Q. **What is the e.m.f. equation of the alternator?**

Ans. $E = 2k_f \cdot k_d \cdot k_c \cdot \phi \cdot f \cdot Z$ volts

where E = e.m.f. generated in volts

k_f = form factor i.e. 1.11

k_d = distribution factor

k_c = coil span factor

ϕ = flux in Wb

f = frequency in c/s

Z = number of conductors

Q. **What is the frequency of generation if an alternator is running at 1500 r.p.m. having four poles?**

Ans. 50 c/s.

Q. **How the frequency can be calculated?**

Ans. $f = \dfrac{P \times N}{120}$.

where f = frequency of generation in c/s

N = speed in r.p.m.

P = number of poles.

Q. **An alternator having four poles is running at 1800 rpm. What will be the frequency of induced emf?**

Ans. 60 c/s.

Q. **A 25 c/s alternator is running at 750 rpm, can you tell how many poles does it has?**

Ans. Four.

Q. **An alternator running at 1500 rpm is producing 450 V, 50 c/s supply, now if the speed is reduced to 1200 rpm, how much voltage it will produce?**

Ans. It will give 360 V.

Q. **But what about the frequency?**

Ans. The frequency of the generated emf will drop down because the frequency and speed both are directly proportional to each other.

Q. **A 10 pole alternator will run at what speed for providing 50 c/s supply?**

Ans. 600 rpm.

Q. **An 84 poles hydroturbine is running to generate 50 c/s supply, what is the speed?**

Ans. 71.42 rpm approximately.

Q. **If the speed of an alternator is kept constant and the field current is changed, what will be the effects on frequency of induced emf and magnitude of induced emf?**

Ans. The frequency of generation will be same but the emf induced will be changed correspondingly.

Q. **What will happen if the field current is kept constant and the speed of alternator is increased?**

Ans. The emf induced will be continuously increased but primarily the frequency of emf will increase which will be of great concern. Hence this procedure is not encouraged.

Q. **Which generator is generally used for excitor and how the induced voltage is fed to the alternator?**

Ans. D.C. shunt generator and the voltage is fed through the leads and terminals.

Q. **How the frequency is governed?**

Ans. By governing the speed of prime mover.

Q. **On what factors does the frequency of an alternator depend?**

Ans. Poles and speed.

Q. **What is the coil span factor?**

Ans. It is defined as the ratio of the resultant e.m.f. to the algebraic sum of the voltage induced in both the coil sides.

It is denoted by k_C which is equal to cos $\beta/2$

where $\beta = 180/3n$, where n = number of slots/poles/phase.

Q. **What is the distribution factor?**

Ans. It is denoted by k_d and is equal to

$$k_d = \frac{\sin \frac{n\alpha}{2}}{n \sin \frac{\alpha}{2}}$$

Q. **What are n and α in the distribution factor?**

Ans. n = the number of slots per pole per phase, and α = the angular displacement between the slots or degree per slot.

Q. **Will the e.m.f. induced in both coil sides of an alternator be same?**

Ans. No, it will not be same in case of short pitched and long pitched windings and may be same in case of whole coil winding i.e. full pitched winding.

Q. **Why in both the short pitched and long pitched windings, the e.m.f. induced in both coil sides is not same?**

Ans. Because the coil sides are not experiencing the same magnitude of the flux at that speed. As the rate of change of flux is responsible for the production of e.m.f. so here speed is naturally same but the magnitude of the flux linking both the coil sides is not of same magnitude, so the e.m.f. induced in both coil sides will not be same.

Q. **What will happen to the e.m.f. if the frequency is doubled?**

Ans. The generated e.m.f. will be doubled.

Q. **If you increase the number of conductors what effect will it make on the frequency of generation?**

Ans. No effect because frequency depends upon speed and number of poles.

Q. **What is done to increase the generated e.m.f. of an alternator?**

Ans. Increase flux by regulating the field current.

Q. **If there are only two sliprings on the shaft of a three phase alternator, which type of alternator is it?**

Ans. Rotating field type.

OPERATION

Q. **What will you do to increase the output of the alternator? Will you increase the speed, field excitation or fuel?**

Ans. Increase the fuel.

Q. **To increase the field excitation voltage, the rheostat is connected in which circuit?**

Ans. Field circuit of shunt generator.

Q. **In which unit the alternators are rated?**

Ans. kVA or MVA.

Q. **Why the rating is in kVA not in kW?**

Ans. Because the power factor of load is not constant.

Q. **If the power factor of the load is low, how will it effect the winding of the alternator?**

Ans. The winding will carry more current for the same real power at low power factor.

Q. **Name the instrument used for power factor controlling in factories?**

Ans. Minimum power factor demand indicator.

Q. **Why the voltage of an alternator falls down with the increasing of the load?**

Ans. Because of the impedance drop and armature reaction.

Q. **What is the impedance drop?**

Ans. The voltage drop because of the load current in armature circuit and the impedance of the winding.

Q. **What do you mean by the synchronous impedance of the alternator?**

Ans. It is the opposition because of the armature effective resistance and synchronous reactance of the winding.

Q. **What do you mean by the effective armature resistance?**

Ans. It is the resistance offered by the armature winding and in A.C. it is increased about 60% because of the skin effect and is known as the effective armature resistance.

Q. **What is leakage reactance?**

Ans. The flux produced because of the current in armature conductors does not completely cross the air gap but complete its path through the stator, such flux is known as flux leakage and due to this the armature winding is supposed to possess a reactance which is known as leakage reactance.

Q. **What is synchronous reactance?**

Ans. The sum of the leakage reactance and fictitious reactance is known as synchronous reactance. It is denoted by X_s and unit is ohm.

Q. **The voltage drop in an alternator is because of which factors?**

Ans. These are load current, armature resistance, leakage reactance and the armature reaction.

Q. **The magnitude of the voltage drop is mainly dependent upon ......?**

Ans. Armature current.

Q. **Can you name the factor which is responsible for the behaviour of this voltage drop.**

Ans. The power factor.

ARMATURE REACTION

Q. **Define armature reaction?**

Ans. It is the effect of the flux produced by the armature conductors on the distribution of flux under the main poles.

Q. **How many types of load is possible in case of alternator?**

Ans. Resistive, inductive, capacitive and mixed.

Q. **What is the effect of the purely resistive load?**

Ans. The main field is distorted and weakened.

Q. **How will you experience about the effect of the purely inductive load?**

Ans. It is purely demagnetising.

Q. **Is there any kind of load which has the magnetising characteristic?**

Ans. Yes, purely capacitive load.

Q. **What is the power factor in case of purely resistive, inductive and capacitive loads?**

Ans. Unity, zero lagging and zero leading, respectively.

VOLTAGE REGULATION

Q. **What is voltage regulation?**

Ans. It is defined as the rise in voltage shown in the percentage of full load voltage, when the full load is thrown off.

Q. **In which unit the regulation is expressed?**

Ans. It is a ratio hence no unit.

Q. **In case of purely resistive load, will there be any difference between the voltage induced (e.m.f.) and the terminal voltage (V).**

Ans. Yes, the e.m.f. will be more than the potential difference V and e.m.f. will be leading the terminal voltage by some angle.

Q. **Can the e.m.f. induced be less than the terminal voltage?**

Ans. Yes, in case of purely capacitive load.

Q. **What is the formula for finding out the generated e.m.f. at unity power factor, lagging power factor and leading power factor?**

Ans. $E = \sqrt{(V\cos\phi + I_aR_a)^2 + (V\sin\phi \pm I_aX_s)^2}$

where E = induced voltage in volts

V = terminal voltage in volts

ϕ = angular displacement

I_a = armature current in amperes

R_a = effective armature resistance in ohms

X_s = synchronous reactance in ohms.

Q. **Why in the portion ($V \sin \phi \pm I_a X_s$), sign ± is used?**

Ans. It is a common formula, + sign is used for lagging and resistive loads but – sign is used for leading power factor.

Q. **How the power factor can be improved?**

Ans. By using the synchronous capacitor and bank of capacitors.

Q. **What is the synchronous capacitor?**

Ans. The overexcited synchronous motor behaves like a synchronous capacitor.

Q. **How actually the power factor correction is made?**

Ans. Just by reducing the reactive component which will certainly improve the power factor.

SYNCHRONIZING OF ALTERNATORS

Q. **If the load on a power station is increased, what will be done to meet the increasing demand of the load?**

Ans. Other alternator is put in parallel to meet the increasing demand, that is known as synchronizing of alternator.

Q. **What is the utility of synchronizing the alternators?**

Ans. Better efficiency, periodic overhauling, general repair, continuation of electric supply in case of any fault in the generating station etc.

Q. **What are the conditions to be fulfilled for ideal synchronizing?**

Ans.

1. The voltage of incoming machine must be same as that of busbar voltage.
2. The phase sequence must be same.
3. The frequency must be same.

Q. **How all these conditions are satisfied?**

Ans. By means of voltmeter, phase sequence indicator and frequency-meter, respectively.

Q. **What are the different methods of synchronizing the alternators?**

Ans. Dark lamp method, bright lamp method, dark and bright lamp method and synchroscope method.

Q. **What should be the indications of lamps in case of dark lamp method and how these are connected?**

Ans. Lamps should be dark and these are connected across the same phases.

Q. **What are the drawbacks of this method?**

Ans. In this method it is difficult to judge the zero value of voltage because even at less voltage the lamp may not glow and may cause some error.

Q. **How the lamps are connected in bright lamp method?**

Ans. Across the opposite phases.

Q. **Which method is generally preferred for synchronizing the alternator?**

Ans. Dark and bright lamp method.

Q. **How the lamps are connected in case of dark and bright lamp method?**

Ans. Dark lamp across the same phase and bright lamps across the opposite phases.

Q. **If the voltage of generation and frequency is alright, what conditions of lamps will be suitable for synchronizing the alternators?**

Ans. The dark lamp must be dark and bright lamps be bright for correct synchronising.

Q. **If the required condition is not achieved what should be done?**

Ans. It can be because of wrong connections. Change it and observe that the same conditions are achieved.

Q. **Name the instrument used for synchronizing the alternators?**

Ans. Synchroscope.

Q. **On what basis does the synchroscope work?**

Ans. On the production of the rotating magnetic field.

Q. **How many windings are there and how these are connected?**

Ans. Two windings, both are fed, one from the incoming machine and other from the busbar or already running machine.

Q. **How the correct position is achieved?**

Ans. There is a marking on the disc of the synchroscope, if it stops and mark is standstill then correct position is obtained.

Q. **If the disc rotates in either direction what does it indicate?**

Ans. The machines are not in synchronizing state. The frequencies are not same.

Q. **What do you mean by the floating of alternator on busbar?**

Ans. While the alternator is neither delivering nor drawing any power from the busbar, the position is known as floating on the busbar.

Q. **What is to be adjusted to share the load?**

Ans. Increase the field excitation of the machine to be loaded.

Q. **If an alternator is to be shut down, what should be done?**

Ans. First of all reduce the load current to minimum possible value say zero by decreasing the field excitation of the machine, then off the main switch or circuit breaker and finally stop the prime mover.

VOLTAGE REGULATION AND REGULATORS

Q. **How the voltage of an alternator is controlled?**

Ans. By manual controlling and by automatic controlling.

Q. **What are the types of automatic regulator?**
Ans. Vibrating type and sliding type.

Q. **Name the vibrating type automatic regulator?**
Ans. Tirill regulator.

Q. **Can you tell any example of the sliding contact type automatic regulator?**
Ans. The silver stud regulator and Brown Bovery regulator.

Q. **Can you tell how actually all types of regulator control the voltage?**
Ans. By controlling the field excitation of the alternator.

Q. **What is open circuit characteristic?**
Ans. The relationship between the voltage induced and the field excitation is known as open circuit characteristic.

Q. **What do you understand by the short circuit characteristic?**
Ans. It is the relationship between the short circuit current and the field current of the alternator.

Q. **In S.C.C. how much short circuit current is permissible?**
Ans. It is about 1.5 to 2 times the full load current but speed is kept constant.

Q. **What are the different methods for finding out the voltage regulation?**
Ans.
1. The synchronous impedance method or EMF method.
2. The ampere-turn or m.m.f. method.
3. Potier method or zero power factor method.

Q. **What are the switchboard equipment for an alternator?**
Ans. Circuit breaker, voltmeter M.I. type (three in number), voltmeter M.C. type (one number), ammeter M.I. type (three in number), ammeter M.C. type (one number), energy meter (one number), power factor meter, frequency meter, wattmeter, synchroscope, field regulator, speedometer (one each).

Q. **What are the possible data you can obtain from the nameplate of an alternator?**
Ans. Capacity in kVA, voltage in volts, kW at specific power factor, frequency in c/s, current in amps, speed in r.p.m., number of phases, number of poles, current in amps., field voltage in volts, temperature rise, type of insulation and serial number of the set etc.

Q. **What do you mean by hunting?**
Ans. It is defined as the oscillation due to sudden loading and switching off of the load.

Q. **What arrangement is done to avoid these huntings?**
Ans. Damper winding is provided.

Q. **Where this winding is housed?**

Ans. In the pole shoes near pole faces and are short circuited as that of the squirrel cage winding in induction motors.

LOSSES AND EFFICIENCY

Q. **What are constant losses?**

Ans. The losses which are constant from no load to full load are known as constant losses.

Q. **What is efficiency?**

Ans. $\% \eta = \frac{\text{Output}}{\text{Input}} \times 100$

Q. **Is cooling necessary in the alternators?**

Ans. Yes, because of load current heat is developed which is certainly required to be dissipated.

Q. **What provision is made for normal cooling?**

Ans. A fan is provided to circulate air, which helps in dissipating the heat developed.

Q. **In big alternators which cooling media is generally provided?**

Ans. Hydrogen cooling.

Q. **Is there any special attention provided for rotor and stator construction of the big alternator for cooling?**

Ans. Yes, some ventilating ducts are provided, horizontally and vertically for better and effective coolant circulation even through the cores.

Q. **If any alternator is down for overhauling, will the shaft of the alternator be left to standstill?**

Ans. No, simultaneously a small motor generally comes in operation and will keep the rotor running at low speed, so that the thermal stresses may not cause any defect or deformation in the shaft or rotor of the alternator.

OBJECTIVE TYPE MULTIPLE CHOICE QUESTIONS

1. Three things are essential for producing an e.m.f. in an alternator ......
   (a) conductor, magnetic field and yoke
   (b) magnetic field, conductor and motion
   (c) conductor, load and motion
   (d) insulation, relative motion and body.
2. Which rotor is most suitable for a turbo alternator ......
   (a) salient pole type
   (b) smooth cylindrical type

(c) both salient or smooth cylindrical type
(d) None of the a, b, c.

3. The e.m.f. equation of the alternator is ......
(a) $E = 4.44\ \phi_m \cdot f \cdot N$ volts
(b) $E = \dfrac{\phi \cdot Z \cdot N \cdot P}{60 \cdot a}$ volts
(c) $E = 2\ k_d \cdot k_f \cdot k_c \cdot \phi_m \cdot f \cdot Z$ volts
(d) $E = 4.44\ k_d \cdot k_f \cdot k_c \cdot \phi_m \cdot f \cdot N$ volts.

4. The distribution factor (k_d) can be given as ......
(a) $\dfrac{\sin \frac{n\alpha}{2}}{n \sin \frac{\alpha}{2}}$ (b) $\dfrac{\sin \frac{n\alpha}{2}}{\alpha \sin \frac{n}{2}}$ (c) $\dfrac{\sin \frac{n\alpha}{2}}{n\alpha \sin \theta}$ (d) $\dfrac{\sin n \times \frac{\alpha}{2}}{\theta \cdot \alpha \cdot \sin \frac{\alpha}{2}}$

5. The coil span factor (k_c) is given as ......
(a) $\cos \dfrac{\beta}{2}$ (b) $\cos (180 - \beta)$
(c) $\dfrac{\sin \frac{n\alpha}{2}}{n \sin \frac{\alpha}{2}}$ (d) $\cos \left(\dfrac{\beta - \alpha}{2}\right)$

6. In rotating armature type alternator supply for load is taken from ......
(a) stator terminals
(b) rotor terminals (sliprings)
(c) partly from slipring and partly from stator winding
(d) from the shaft.

7. If the input to the prime mover of an alternator is kept constant but the excitation is changed, then the ......
(a) reactive component of the output is changed
(b) active component of the output is changed
(c) the power factor of the load is unaffected
(d) the frequency of the induced e.m.f. changes.

8. The armature reaction in case of purely capacitive load is ......
(a) demagnetising (b) magnetising
(c) cross-magnetising (d) None of the a, b, c

9. The rating of the alternator is given in ......
(a) kW (b) kVA (c) kVAR (d) kV.

10. Synchroscope is used to ......
(a) measure the resistance of the windings
(b) synchronize the alternators

(c) synchronize the field and armature windings
(d) measure the percentage reactance drops.

11. The power factor of the alternator depends upon ......
(a) load current (b) load nature
(c) rotor speed (d) core losses.

12. The turbo alternator runs at very ...... speed and ...... type rotor is used in it.
(a) low; salient pole (b) high; smooth cylindrical pole
(c) high; salient pole (d) low; smooth cylindrical pole

13. The hydro turbine runs at ...... speed and ...... type rotors are used in it.
(a) low; salient pole (b) high; smooth cylindrical pole
(c) high; salient pole (d) low; smooth cylindrical pole.

14. Generally the field excitor is mounted on the same shaft, it is a ......
(a) D.C. series generator (b) D.C. compound generator
(c) D.C. shunt generator (d) None of the above.

15. When two alternators are supplying a load, these are generally said to be ......
(a) paralleled (b) synchronized
(c) in series (d) out of phase.

16. When two alternators are synchronized and by the way, the prime mover of one alternator fails, then this alternator will ......
(a) burn out (b) work as an induction motor
(c) work as a synchronous motor
(d) stop.

17. An alternator is termed as overexcited if ......
(a) it is supplying a unity power factor load
(b) it is supplying a lagging power factor load
(c) it is supplying a leading power factor load
(d) it is supplying a mixed lagging power factor load.

18. For synchronizing two alternators, the following conditions must be fulfilled.
(a) The voltage must be same.
(b) The frequency must be same.
(c) The phase sequence must be same.
(d) The voltage, frequency and phase sequence must be same.

19. For the same power capacity the higher voltage alternator will be ......
(a) bigger in size (b) smaller in size
(c) less costly (d) costly.

20. If an alternator is running and operating on leading power factor, the voltage regulation will be ......
    (a) negative (b) positive
    (c) equal to zero (d) None of the above.
21. Alternators above 50 MW are provided with ...... cooling.
    (a) air blast (b) natural air
    (c) hydrogen (d) water.
22. The advantages of hydrogen cooling are ......
    (a) increased efficiency and rating
    (b) increased rating and life
    (c) increased efficiency, rating life and eliminate fire hazards
    (d) increased life, rating, fire hazard elimination and less noise.
23. The speed of a six pole alternator producing 250 V, 50 c/s is ......
    (a) 1500 rpm (b) 500 rpm (c) 330 rpm (d) 1000 rpm.
24. The three phase alternator stationary field type has ......
    (a) two sliprings (b) four sliprings
    (c) three sliprings (d) one slipring.
25. High speed alternators are ......
    (a) smooth cylindrical rotor type
    (b) salient pole type
    (c) revolving field and salient rotor type
    (d) None of the above.

CHAPTER 17

Transformer

INTRODUCTION

Q. **What is a transformer?**

Ans. The transformer is a static device which is used to transfers the electrical power from one volt to another volt, without changing the frequency.

Q. **What do you understand by step up and step down transformers?**

Ans. The transformer which changes high voltage into low voltage is known as step down transformer and the transformer which changes low voltage into high voltage is known as step up transformer.

Q. **Why there is no change in the frequency of transformation of the voltage in transformer?**

Ans. The flux which is the base of inducing voltage in the transformer is same for both the windings of the transformer or say the flux in the cores has got definite frequency and it will be same on the primary and secondary sides hence the question of changing the frequency does not arise.

Q. **What is the working principle of the transformer?**

Ans. The transformer works on the principle of "Faraday's laws of electromagnetic induction." According to that, whenever a conductor links or associates with the changing flux an e.m.f. is induced in the conductor, this e.m.f. is also proportional to the rate of change of flux and the number of conductors.

Q. **Which type of induction takes place in the two winding transformer?**

Ans. Mutual induction.

Q. **What are the different types of windings?**

Ans. Low tension (L.T.) and high tension (H.T.)

Q. **If supply is given to (a) L.T. winding and load is connected on H.T.**

(b) H.T. winding and load connected on L.T. winding, name the type of the transformer?

Ans. (a) Step up transformer.
(b) Step down transformer.

Q. **Name the winding which is connected across the supply?**

Ans. Primary winding.

Q. **Where load is connected?**

Ans. On the secondary winding.

Q. **If the winding has more number of turns the winding is called H.T. or L.T.?**

Ans. H.T.

Q. **Is there any relationship between the flux linkage and e.m.f. induced?**

Ans. Yes, e.m.f. induced is proportional to flux linkage.

CONSTRUCTION

Q. **What are the different parts of the transformer?**

Ans. Cores, windings, oil, tank, conservator and breather, and cooling arrangement.

Q. **What type of cores are used?**

Ans. Core type, shell type and berry type.

Q. **What is the function of the cores?**

Ans. The main function is to offer an easy path for the lines of force and to reduce the eddy current losses etc.

Q. **How many magnetic paths are there in core type cores?**

Ans. One path.

Q. **Name the cores which have two magnetic paths?**

Ans. Shell type.

Q. **If the distributed magnetic circuits are required which cores are used?**

Ans. Berry type construction.

Q. **Which material is used for cores?**

Ans. Laminations of silicon steel.

Q. **Why a laminated solid block is used instead of a solid homogeneous block?**

Ans. To reduce the eddy current losses.

Q. **What are the types of winding according to the construction?**

Ans. Sandwich type and cylindrical type.

Q. **According to the voltage, the windings are named as?**

Ans. H.T. and L.T. winding.

Q. **What are the cylindrical type winding?**

Ans. In this type of coil, the length of the coil is equal to the length of the core limb. The primary and secondary winding may be placed one over the other.

Q. **What is the sandwich type winding?**

Ans. If the high tension and low tension windings are placed one over the other like the sandwiches limb-length-wise, this type is known as sandwich type winding.

Q. **While selecting the method of placement of the winding, which winding is placed near the core, in sandwich type winding?**

Ans. L.T. and then H.T.

Q. **Why L.T. winding is placed near the cores?**

Ans. To reduce the dielectric strength of the dielectric or insulation provided. In this case the total dielectric strength of the insulation will be less than if H.T. would be near the cores.

Q. **Is there any insulation between the laminations?**

Ans. Yes, insulating varnish or paper.

Q. **What do you mean by the limb?**

Ans. It is vertical portion where the windings are placed.

Q. **How many limbs are there in (i) U shape core (ii) E shape cores?**

Ans. (i) Two only. (ii) three only.

Q. **How the central limb and side limbs of the E shape are designed?**

Ans. Central limb has double the width than the side limbs.

Q. **Why double width?**

Ans. As the flux has two magnetic paths and total flux will pass through the central limb.

Q. **What is the transformer oil?**

Ans. It is a mineral oil. It is filled in the tank of the transformer.

Q. **What are the functions of the transformer oil?**

Ans. It is used for heat dissipation, insulation and spark quenching medium also.

Q. **In which units the dielectric strength of the transformer oil is measured?**

Ans. kV/mm.

Q. **What are the factors that affect the dielectric strength of the transformer oil?**

Ans. The moisture, dust and temperature.

Q. **Will the moisture increase or decrease the dielectric strength?**

Ans. Decrease.

Q. **How the moisture contained is removed or dehydration is performed?**

Ans. It is cleaned and passed through the dehydration plant. The moisture is removed and dehydration increases the dielectric strength of the transformer oil.

Q. **How the tank of a transformer is made?**

Ans. The tank of small transformer is made of the iron sheet welded and for big transformer it is made from plain boiler plates.

Q. **What is conservator?**

Ans. Conservator is a small tank mounted on the top of the transformer tank. The transformer oil is filled in it. There is a provision to check the colour and level of the oil. The conservators are required to bear the expansion and contraction of the oil. It is generally of cylindrical construction.

Q. **What is the breather?**

Ans. It is a pipe attached with the conservator. It contains the pot at its extreme end which has $CaCl_2$ or silica gel to extract the moisture from the atmospheric air.

Q. **Why it is necessary to have $CaCl_2$ or silica gel in the breather?**

Ans. While temperature of the oil increases, the oil expands causing more volume to be occupied, thus releasing the gas or air to external atmosphere through the breather. Now when temperature decreases, the volume decreases causing an empty space, which is refilled by the air from the atmosphere, sucked via breather. This air may have the moisture which will effect the dielectric strength of the oil, so to extract moisture $CaCl_2$ or silica gel is placed.

Q. **Is there any arrangement provided to check the temperature of the oil?**

Ans. Yes, the temperature gauge.

Q. **Can you fill the conservator completely with oil and if not why?**

Ans. It should never be completely filled because with the increase in temperature, the oil expends which may cause serious trouble for example tank may burst out.

Q. **What is the flash point of the transformer oil?**

Ans. The temperature at which the vapours above an oil surface ignites is known as flash point. For safety reasons the oil not less than 160°C flash point is used in tank.

Q. **What do you understand by sludging in the transformer oil?**

Ans. It is an important characteristic of the oil, which has to be watched very carefully. Sludging means the slow formation of semisolid hydrocarbons on the windings and tank walls.

Q. **Why corrugated sheets are provided?**

Ans. To increase the heat dissipating area of the tank.

Q. **Where actually the heat is developed, in the cores or windings?**

Ans. In both, in the windings because of the current flowing, in the cores because of eddy currents.

Q. **What are the different methods of cooling?**

Ans.
1. Natural air cooling.
2. Oil immersed natural air cooling.
3. Oil immersed forced water cooling.
4. Oil immersed forced air cooling.
5. Air blast cooling.

Q. **What is meant by the air blast cooling?**

Ans. In this case the stream of the forced air is forced to the windings and cores by means of the blowers or high speed fans. The air passes from bottom to top.

Q. **Why forced water cooling is not preferred?**

Ans. Because, if by chance, the tube leaks, the water will come out in the oil and will decrease its dielectric strength.

Q. **Upto 500 kVA. which type of cooling is better?**

Ans. Oil immersed natural air cooling.

Q. **What are the different methods of increasing the dissipating area for better cooling?**

Ans. By providing :
1. The corrugated sheets.
2. Fins welded vertically to the sides of tank.
3. Elliptical or round tubes.
4. Auxiliary radiator tanks.

Q. **What is the Buchhol's relay?**

Ans. It is the protective relay which is meant for the protection of oil immersed transformed from insulation failure, cores excess heating or any other internal faults which cause heating of the oil beyond the limit and maintaining transformer oil level etc.

Q. **Is there any possibility of overflowing the transformer oil?**

Ans. No, a relay is also provided to alarm about the level.

Q. **Is there any protection from the gases or fumes in the tank?**

Ans. Yes, Buchhol's relay.

E.M.F. EQUATION

Q. **What is the e.m.f. equation of the transformer?**

Ans. $E = 4.44\phi_m \cdot f \cdot N$ volts.

Q. **Tell the e.m.f. equation in terms of flux density?**

Ans. $E = 4.44 B_m \cdot A \cdot f \cdot N$ volts.

Q. **How the number of turns of primary winding is calculated if the voltage, frequency and flux are given?**

Ans. $N_P = \dfrac{E}{4.44 \phi_m \cdot f}$

Q. **What is the voltage transformation ratio?**

Ans. The voltage transformation ratio is the ratio of the secondary voltage to primary voltage.

i.e., $\dfrac{V_s}{V_p}$

Q. **What are the current and turn transformation ratios?**

Ans. $\dfrac{N_s}{N_p}$ is the turn transformation ratio.

$\dfrac{I_p}{I_s}$ is the current transformation ratio.

Q. **For an ideal transformer, what is the condition?**

Ans. $V_p \cdot I_p = V_s \cdot I_s$

Q. **Is there any relationship between the voltage, turns and current transformation ratios?**

Ans. Yes. $\dfrac{V_s}{V_p} = \dfrac{N_s}{N_p} = \dfrac{I_p}{I_s} = k.$

OPERATION

Q. **What do you mean by the no load current of the transformer?**

Ans. When the primary winding is connected across the supply and secondary winding is open-circuited, then the transformer is said to be on no load and current taken is called the no load current of the transformer.

Q. **What are the no load current's components?**

Ans. The wattful component i.e. $I_0 \cos \phi_0$ and the wattless or magnetising component i.e. $I_0 \sin \phi_0$.

Q. **What is the wattful component of no load current?**

Ans. The current $I_0 \cos \phi_0$ is used to overcome the hysteresis and eddy current losses, and is known as wattful component.

Q. **What is wattless or magnetising component of the current?**

Ans. It is the current which is used to produce the magnetic flux in the cores and is known as the magnetising component of the no load current, i.e. $I_\mu = I_0 \sin \phi_0$.

Q. **What is the angular displacement between the Iω and Iμ currents?**

Ans. 90°.

Q. **If the I_ω and I_μ are known, can you tell the formula for calculating the no load current?**

Ans. $I_0^2 = I_\omega^2 + I_\mu^2$.

Q. **If B_m, I and N_p are given can you give the formula for finding out the magnetising current?**

Ans. $I_\mu = \dfrac{B_m \cdot I}{\mu_0 \cdot \mu_r \cdot N_p}$ amp.

Q. **What is the utility of the magnetising current? Can you imagine the transformer without the magnetising source?**

Ans. The magnetising current is used to produce the magnetic flux in the cores. The flux is the base of the magnetic induction and without induction the secondary cannot induce any voltage, so without the magnetising source the transformer has no meaning.

Q. **When do we call a transformer of load?**

Ans. When the secondary winding is connected to any load, the transformer is said to be on load.

Q. **Will the current in primary winding and secondary winding be same?**

Ans. No, the primary winding current is the sum of the no load current and secondary transferred current on primary side.

Q. **What about the power factor on primary and secondary side of a transformer?**

Ans. It is not same but generally taken as same.

Q. **Which side power factor is more?**

Ans. Generally primary side, but for all the practical purposes the power factor on primary and secondary windings is taken as same.

Q. **Which type of load is put on the transformer?**

Ans. All type, i.e., resistive, capacitive and inductive and mixed too.

Q. **What do you mean by the resistive drop and reactance drop?**

Ans. The drops in the resistance of the winding is known as resistive drop, i.e. $I \cdot R$ volts and if the drop is taken in reactance of the winding, $I \cdot X$ volts is known as reactive drop.

Q. **Generally which drop resistive or reactive is more in value?**

Ans. Reactive voltage drop.

VOLTAGE REGULATION

Q. **What is the voltage regulation?**

Ans. The voltage regulation is defined as the rise in voltage when the full load is thrown off shown in the percentage of no load voltage, i.e. $$\% \text{ Regulation} = \frac{V_0 - V_L}{V_0} \times 100$$

Q. **What are the methods for finding out the voltage regulation?**

Ans. 1. By actual loading.
2. By calculations.

Q. **If V_2, R_2, X_2, I_2 are given how the voltage regulation is calculated?**

Ans. $$\% \text{ Regulation} = \frac{I_2R_2 \cos \phi + I_2X_2 \sin \phi}{V_2} \times 100.$$

TAP CHANGING

Q. **What do you understand by tap changing?**

Ans. Whenever the load is put on the transformer some voltage drops are inherent, but it is essential to keep the secondary voltage approximately constant. For that reason, the transformation ratio is changed, that changing from one tap to another is termed as tap changing.

Q. **Generally which tap changing is preferred step up or step down?**

Ans. Both as required.

Q. **What are the classifications of tap changing?**

Ans. 1. On load tap changing.
2. No load tap changing.

Q. **What are the no load and on load tap changings?**

Ans. If the tap changing is done, when there is no load on the secondary of the transformer it is called no load tap changing and if there is load on the secondary and tap changing is done, then it is known as on load tap changing.

Q. **What are the methods of no load tap changing?**

Ans. Manual and automatic.

Q. **What type of regulators are used for no load changing?**

Ans. Sliding contact type and face plate type. In these cases the tappings are brought out.

Q. **Give some idea about the sliding contact type and face plate type regulators?**

Ans. *Sliding contact type :* In this case a number of tappings are brought out and contacts are slided manually.

Face plate type : In this case the terminals are brought out and a D.C. motor is energised with the movement of the arm, the contacts are formed and transformation ratio is changed. It is generally having a sector or quadrant type shape.

Q. **What are the methods of on load tap changing?**

Ans. It is done by the following methods :

1. Parallel winding.
2. Preventive reactor.
3. Regulation transformer.
4. Tertiary winding.

Q. **Why in case preventive regulator and parallel winding regulators the tappings are taken from the centre of the winding?**

Ans. To avoid any breaking or cause of any discontinuity in combined circuit. In reactance two contacts of the reactor are joined to the same contact of the tappings, so that the current divides in the same quantity and that too in opposite direction to cancel out the reactance voltage. Thus smooth and undisturbed changing is obtained.

LOSSES AND EFFICIENCY

Q. **What types of losses are there in a transformer?**

Ans. Copper losses, iron losses or stray losses.

Q. **What are copper losses in a transformer?**

Ans. These are the losses in the primary and secondary windings because of the resistive component. These are I^2R watts. These are variable losses.

Q. **Why we call the iron losses as constant losses?**

Ans. These losses are constant from no load to full load. These losses are because of the magnetic flux, which remains constant throughout the working of the transformer.

Q. **What is the efficiency?**

Ans. The efficiency is the ratio of output to input shown in percentage.

Q. **What is the all day efficiency?**

Ans. It is also the ratio of output to input in a fixed time shown in percentage but the duration is taken 24 hrs. So

$$\% \text{ All day efficiency} = \frac{\text{kWh output in 24 hrs}}{\text{kWh input in 24 hrs}} \times 100.$$

TESTING

Q. **How core losses are found out?**

Ans. By means of open circuit test.

Q. **In case of open circuit test, which losses are shown by the reading of wattmeter?**

Ans. Iron losses.

Q. **In open circuit test, how the secondary winding is connected?**

Ans. It is left open.

Q. **Generally which side L.T. or H.T. is left open?**

Ans. H.T.

Q. **Why H.T. is opened?**

Ans. It is because metering on H.T. winding is somewhat difficult.

Q. **Why the reading shown by wattmeter is core losses?**

Ans. Because the current flowing in the winding is very small in comparison to the full load current, but the cores of the transformer are fully energised i.e., having the normal working flux.

Q. **Why short circuit test is carried out?**

Ans. To find out the copper losses.

Q. **Which side is short circuited?**

Ans. L.T. side.

Q. **Cay you short circuit it through an ammeter?**

Ans. Yes, there is no problem but the ammeter should be suitably selected.

Q. **Why the losses shown in short circuit test are considered as the copper losses?**

Ans. Because at that time the voltage fed is small enough but current in the windings is full load or as required, which causes the copper losses. Though the iron losses are also shown, but these are very small in comparison to the actual iron losses of transformer.

Q. **Can you name a single test, which can give both copper losses and core losses in one test?**

Ans. Sumpner test.

Q. **What is the unique characteristic of the Sumpner test.**

Ans. In this test the transformer is loaded without consuming much power for several hours. The single test can show the copper losses, core losses and temperature rise.

Q. **How the primary windings of both transformers are connected?**

Ans. In parallel across the supply.

Q. **The secondary of both the transformers are connected in parallel but in that case there cannot be any circulating current. How the current is made to circulate in the secondaries?**

Ans. By introducing an injucting transformer to injuct some voltage in the secondary circuit.

Q. **How that injucting transformer circulates the current?**

Ans. As both secondaries are connected in parallel so there is no resultant voltage working across, with the help of the injucting transformer, we are injucting some voltage in the circuit which depending upon their impedances, circulates the current in the secondary and the transformer is loaded.

Q. **Is there any instrument connected in the secondary winding?**

Ans. Yes, the wattmeter and the ammeter can also be connected.

Q. **What this wattmeter shows?**

Ans. The copper losses.

Q. **While injucting the voltage, a circulating current starts flowing in the secondaries, it will also cause some current in the primary circuit, it is going to effect the line current or wattmeter which is connected in primary circuit?**

Ans. No, the circulating current will only flow through the windings and their leads, but will not contribute anything to the wattmeters.

Q. **Is it necessary that the injucting voltage must have the same frequency as that of the supply voltage on which primaries are connected?**

Ans. Not necessary.

Q. **What tests should be performed before commencing the transformer?**

Ans. Open circuit test, short circuit test, phasing out test, polarity test, D.C. resistance test, voltage ratio test, magnetising current test, core losses test, efficiency test, temperature rise test, insulation test, impulse test and oil test also.

Q. **Why an impulse test is performed?**

Ans. This test is performed to determine the ability of transformer to withstand the effects of high unidirectional voltages resembling to the sky lightning or surge voltage.

Q. **What is the insulation test?**

Ans. This test is carried out to find out the insulation resistance between the windings, between cores and copper.

Q. **How the core test is performed and how much voltage and frequency is kept for testing?**

Ans. In this test, the H.T. winding is left open and L.T. winding is connected to the normal voltage and frequency through ammeter and wattmeter.

PARALLEL OPERATION

Q. **What do you mean by the parallel operation of the transformer?**

Ans. When the increasing demand of the load is not suitably fed by one transformer other transformer is brought into operation to share the load. This is called the parallel operation.

Q. **How these are connected?**

Ans. In parallel.

Q. **What are the conditions to be fulfilled before parallel operation of the single phase transformers?**

Ans. 1. The voltage transformation ratio must be same.

2. The percentage impedance drop should be same.
3. The polarity should be same.

Q. **What are the conditions to be fulfilled for 3 phase transformer for parallel operation?**

Ans. 1. Voltage transformation ratio should be same.
2. Phase sequence should be same.
3. Phase rotation must be same.
4. Percentage impedance drop must be same.

Q. **Can you tell the necessity of parallel operation?**

Ans. 1. To meet the increasing demand of the load.
2. To continue the supply in case of any fault and regular maintenance also.
3. For periodic overhauling of the transformer.
4. For maximum operational efficiency.

Q. **What do you know about the transmission transformer?**

Ans. It is the transformer which changes the voltage from low voltage to high voltage in generating station for onward transmission. These are high capacity transformers.

Q. **What do you know about the distribution transformer?**

Ans. The transformer which is used to distribute the electrical power to the consumers is known as distribution transformer. The primary of the transformer is connected in delta and secondary in star to facilitate for three phase four wires distribution. These are designed to work for 24 hours.

Q. **What do you understand from the power transformer?**

Ans. The transformers which are installed in big industries and are used for any specific and particular purposes. These are designed for maximum efficiency and less duration. These are energised for short period only for the need, i.e. operation.

Q. **How the current is shared in parallel operation?**

Ans. According to their capacity and characteristic of the transformer.

TYPES

Q. **What are the types of the transformer according to the number of phases?**

Ans. Single phase, two phase, three phase, six phase, etc.

Q. **How the transformers are classified according to the core construction?**

Ans. Core type, shell type and distributed core type.

Q. **How many limbs are there in a core type stamping used for three phase transformer?**

Ans. Three of equal length and area.

Q. **How many limbs are there in a stamping used for three phase shell type transformer?**

Ans. Five.

Q. **What types of connections are there for three phase transformer?**

Ans. Star/star, delta/delta, star/delta, delta/star. open delta and scott connections are also possible.

Q. **Where the following connections are generally used : (a) Star/delta, (b) Delta/star?**

Ans. (a) Transmission transformer.
(b) Distribution transformer.

Q. **What are the advantages of using star connections in the distribution transformer?**

Ans. Because for distribution purpose, the three phase four wire system is generally used. The lighting load and power load can be put on the same supply. Power load on three phase and lighting load on one phase and one neutral.

Q. **What is the utility of delta connections?**

Ans. For transmission purpose, where the power is to be transmitted to a far distance, only three lines will solve the purpose. Thus the cost of transmission is saved, secondly in open delta, the supply will be continued but with 60% efficiency only.

Q. **How many ways are generally used for three phase transformation?**

Ans. Single unit of three phase and three units of single phase transformers.

Q. **What are the utilities of using a three phase single transformer?**

Ans. It requires less space, less cost and less maintenance also.

Q. **Is there any disadvantage of using a single unit of three phase?**

Ans. Yes, in case of fault the whole unit is required to be repaired or replaced.

Q. **What are the drawbacks of using three single phase transformers?**

Ans. More space is required, cost is more and the maintenance cost is also more.

Q. **Can a three phase supply be transformed into two phase?**

Ans. Yes, by using Tee or Scott connections.

Q. **What are the uses of Scott connections?**

Ans. In this method, the three phase supply can be converted into two phase supply for welding purpose.

Q. **Which is an ideal transformer?**

Ans. The transformer which has 100% efficiency.

Q. **What is the leakage reactance?**

Ans. Whenever the flux is induced by the winding some flux links with only this particular winding and causes some self-induced voltage in it. This flux is called the leakage flux and the reactance because of this flux is leakage reactance of that particular winding.

Q. **What do you understand by the self-induced voltage in primary winding?**

Ans. Whenever the primary winding is energised it produces some magnetic flux. The flux can be divided into two parts—one, which induces the voltage in the secondary winding, the other which only links with primary winding and causes primary self-induced voltage, which opposes the applied voltage.

Q. **What is the equivalent resistance of the transformer referred to primary winding?**

Ans. $R_{01} = R_1 + \frac{1}{\rho^2} R_2$ ohm.

where R_{01} = equivalent resistance referred to primary winding; R_1 = resistance of primary winding; R_2 = resistance of secondary winding; and ρ is the transformation ratio of the transformer.

Q. **What is the equivalent reactance of the transformer referred to secondary winding?**

Ans. $X_{02} = X_2 + X_1 \rho^2$ ohm.

Q. **In which units the capacity of a transformer is shown?**

Ans. kVA.

Q. **Why it is shown in kVA and not in kW?**

Ans. Because the load power factor is not given. If it is in kW the power factor is fixed, which is not always applicable.

INSTRUMENT TRANSFORMER

Q. **What are the instrument transformers?**

Ans. The current transformer (C.T.) and potential transformer (P.T.) are known as instrument transformers.

Q. **Where C.T. is used and what is the type of this transformer?**

Ans. It is a current transformer and used to measure high currents. It is a step up transformer.

Q. **What is P.T.?**

Ans. It is the potential transformer and used for measuring high voltage. It is a step down transformer.

Q. **Can you tell the amperes for which the C.T. secondary is generally designed?**

Ans. 5 A.

Q. **The secondary of the P.T. is designed for how many voltage?**

Ans. Generally 110 V.

Q. **Can a wattmeter be connected with C.T. and P.T. for measuring power on H.T. line?**

Ans. Yes.

Q. **Is there any change in the core flux of a transformer, under loaded and without load conditions?**

Ans. No, it remains same under all conditions of load.

Q. **Should the secondaries of both C.T. and P.T. be earthed, if yes, why?**

Ans. Yes, the secondaries should be earthed, to protect against electric shock.

Q. **What precautions should be taken while replacing the ammeter of C.T. circuit?**

Ans. The secondary terminals must be short circuited before taking out the ammeter. Replace the meter and open the terminals. It is essential that in no case the secondary of the C.T. be left open.

Q. **The Tong tester is an example of C. T. or P.T.?**

Ans. C.T.

AUTOTRANSFORMER

Q. **What is an autotransformer?**

Ans. The transformer which has only one winding, both primary and secondary are permanently connected electrically, magnetically and mechanically. A portion of the winding works as the primary and other as secondary, such transformer is known as autotransformer.

Q. **What is the working principle of an autotransformer?**

Ans. Self-induction.

Q. **How the secondary voltage is calculated?**

Ans. In a particular transformer, the voltage per turn is constant. So the voltage can be calculated, multiplying the voltage per turn to the number of secondary turns.

Q. **Are the number of turn per voltage constant for a single winding transformer or two winding transformer?**

Ans. Constant for the particular transformer whether it is single phase or three phase, single winding or two winding transformer.

Q. **For maximum efficiency can you tell the transformation ratio of an autotransformer?**

Ans. Yes, it is 3/4.

Q. **What are the advantages of an autotransformer?**

Ans. Copper saving, only one winding is sufficient, light in weight, labour cost is less.

Q. **What are the common uses of an autotransformer?**

Ans. Used as a starter, stabilizer for T.V. and refrigerators, and in long distance line as a booster also.

Q. **What are the disadvantages of autotransformer?**

Ans. There is only one winding, so the secondary portion is also energised and may give shock even when the load is not connected.

Q. **Is the current in both primary and secondary portions same or different?**

Ans. Different, in the common portion the current is the difference of primary and secondary currents.

Q. **Is it advisable to use the same conductor for primary and secondary windings of an autotransformer?**

Ans. Not advisable and will be merely a wastage.

Q. **Is there any copper saving in an autotransformer than two winding transformer?**

Ans. Yes, depending upon the transformation ratio. The better and maximum saving is when $\rho = 3/4$.

Q. **What safety devices are used for a transformer?**

Ans. Air circuit breaker, lightning arrester, relays and earthing. etc.

Q. **If a transformer is connected to D.C. what will happen?**

Ans. It will not induce any voltage (except the rise and decay time) and finally because of less winding resistance it will burnt out.

Q. **What will be the direction of currents in primary and secondary windings/portions?**

Ans. Opposite to each other.

Q. **What is a tertiary winding?**

Ans. The transformers are sometimes designed with three windings, two main windings and third is the tertiary winding, which is used to make a closed mesh connection to stabilize the neutrals of both primary and secondary circuits.

Q. **What do you understand by the transformer bushings?**

Ans. The bushings are provided for insulating purposes and bringing the terminals of the winding out for connections to the external circuit.

Q. **Which harmonic is dangerous and which winding is placed to carry it?**

Ans. The third harmonic and the winding is tertiary winding.

OBJECTIVE TYPE MULTIPLE CHOICE QUESTIONS

1. The transformer is a ...... device.
   (a) static (b) dynamic
   (c) primary (d) secondary
2. The supply is connected to ...... of the transformer.
   (a) secondary winding (b) primary winding
   (c) L.T. (d) H.T.
3. If a transformer changes high voltage to low voltage it is known as ......
   (a) step up transformer (b) step down transformer
   (c) one to one transformer (d) autotransformer.
4. In a power transformer the winding which is nearer to cores is ......
   (a) primary winding (b) secondary winding
   (c) low voltage winding (d) high voltage winding.
5. The transformer efficiency is generally in the range of ......
   (a) 50-60% (b) 60-70% (c) 70-84% (d) 90-98%
6. The transformer changes the voltage level of supply ......
   (a) without changing the power
   (b) without changing the frequency
   (c) without changing the sine wave
   (d) without changing the power, frequency and wave.
7. The open circuit test is performed to find out the ......
   (a) core loss of the transformer
   (b) copper loss of the transformer
   (c) efficiency
   (d) windage losses.
8. The e.m.f. equation of the transformer is given as ......
   (a) $V = 4.44\phi_m \cdot f \cdot N$ volts (b) $V = \dfrac{\phi \cdot Z \cdot N \cdot P}{60 \cdot a}$ volts
   (c) $V = 2.22A_m \cdot f \cdot N$ volts (d) $V = 4B_m \cdot A \cdot f \cdot N$ volts.
9. An ideal transformer has maximum efficiency when ......
   (a) copper losses are equal to iron losses
   (b) copper losses are less than iron losses
   (c) copper losses are more than iron losses
   (d) copper losses and iron losses are equal to windage losses.
10. The purpose of oil in the tank of a transformer is ......
    (a) to insulate the windings
    (b) to dissipate heat developed

(c) to quench the spark, if any
(d) to insulate windings, cores, dissipate heat generated and to quench the spark, if any.

11. Instrument transformers are used to measure ......
(a) current (b) voltage
(c) frequency (d) current and voltage both.

12. In autotransformer the primary of secondary windings are ......
(a) connected mechanically
(b) connected electrically
(c) connected magnetically
(d) connected electrically, mechanically and magnetically.

13. The transformer in which winding is surrounded by the cores is known as ......
(a) shell type (b) core type
(c) tertiary type (d) None of the above a, b, c.

14. ...... transformer is used to measure the heavy load currents.
(a) Current (b) Potential
(c) Autotransformer (d) Ammeter.

15. ...... transformer is used to measure high voltage of A.C.
(a) Heavy current (b) Potential
(c) Autotransformer (d) Voltmeter.

16. The transformer cores are made of laminations of ......
(a) silicon steel (b) annealed copper sheets
(c) aluminium sheets (d) tungsten steel.

17. The laminations of silicone steels are used to reduce the ...... losses
(a) hysteresis (b) eddy current
(c) eddy current and hysteresis (d) magnetic noise.

18. The all day efficiency of a transformer is the ratio of ......
(a) kWh input to kWh output
(b) kWh output to kWh input in 24 hours
(c) output power to input power in 24 hours
(d) output power to input energy in 24 hours

19. Maximum efficiency of a transformer can be achieved if ......
(a) eddy current and hysteresis losses are equal
(b) copper losses and iron losses are equal
(c) the power factor of the load is leading
(d) eddy current losses are equal to copper losses.

20. For satisfactory parallel operation ......
(a) the voltage transformation ratio should be same

(b) the polarity should be same
(c) the percentage impedance drop should be same
(d) the polarity, transformation ratio and percentage impedance drop should be identical.

21. The transformer oil is a ......
    (a) mineral oil
    (b) vegetable oil
    (c) kerosene oil
    (d) coconut oil.
22. A 100 MVA transformer has ...... method of cooling.
    (a) A.N.
    (b) ONAN
    (c) OFAF
    (d) None of the above.
23. The silica gel becomes ...... after absorbing moisture.
    (a) yellow (b) blue (c) brown (d) green.
24. Even if the transformer is fully loaded, the ...... will be zero.
    (a) copper loss
    (b) eddy current loss
    (c) friction loss
    (d) hysteresis loss.
25. The sound generated by a transformer is generally termed as ......
    (a) humming (b) ringing (c) buzzering (d) zooming
26. The primary winding of a transformer is always wound with thick conductor.
    (a) True
    (b) False.
27. The autotransformer has two separate windings wound of different limbs.
    (a) True
    (b) False.
28. The potential transformer is used to measure voltage.
    (a) True
    (b) False.
29. The voltage stabilizer is an autotransformer.
    (a) True
    (b) False.
30. Horn gap arrester is used for filling up of oil.
    (a) True
    (b) False.

CHAPTER 18

A.C. Motors

INTRODUCTION

Q. **What is an A.C. motor?**

Ans. The motor which works on A.C. supply is known as A.C. motor.

Q. **What are the main classifications of A.C. motors?**

Ans. These are divided into :

1. Synchronous motor
2. Asynchronous motor.

Q. **What do you know about the synchronous and asynchronous or non-synchronous motor?**

Ans. The motor which works on synchronous speed is known as the synchronous motor and that which does not run at synchronous speed is known as asynchronous or non-synchronous motor.

Q. **Can you give some names of non-synchronous motors?**

Ans. These are squirrel cage induction motor, slip-ring induction motor or wound rotor motor, double squirrel cage induction motor.

Q. **What is meant by an induction motor?**

Ans. The motor in which supply is not given to the rotor winding, but it is induced in it because of the induction, is known as induction motor.

Q. **What is the working principle of an induction motor?**

Ans. "Whenever a current-carrying conductor is placed in the magnetic field, a torque is developed over the conductor, which tends to rotate.

Q. **When the rotor winding of the motor is not fed, how the current circulates in the rotor conductors?**

Ans. Because of induction some e.m.f. is induced in the rotor bars. The bars are short circuited so the current circulates and torque is developed.

Q. **How the direction of torque developed is found?**

Ans. By means of Fleming's Left Hand Rule.

Q. **What is the Fleming's Left Hand Rule?**

Ans. According to this rule, stretch your left hand in such a ways so that the thumb, first finger and middle fingers are at right angle to each other, then if the first finger indicates the direction of magnetic flux, the middle finger the direction of current in the conductor then thumb will indicate the direction of torque developed on the conductor.

CONSTRUCTION

Q. **What are the constructional parts of the induction motor?**

Ans. These can be divided into two—the stationary parts and rotating parts. These are the body, side covers, bed, eyebolt, terminal box, shaft, rotor, bearings and fan.

Q. **What do you mean by stator and rotor?**

Ans. The stator is the stationary portion of the motor and that which rotates is known as rotor.

Q. **Can you tell the construction of rotor?**

Ans. It is a rotating part. It has so many slots. There are copper bars or aluminium bars placed in the slots. These bars are short-circuited at both the ends. These short circuiting rings are known as end rings.

Q. **What is the function of short circuited bars in the rotor?**

Ans. It provides the current-carrying conductor in the magnetic field.

Q. **How the current circulates in the rotor bars?**

Ans. The e.m.f. induced in the bars when short-circuited by the end-rings causes the current to circulate through the bars.

Q. **Can you name the fashion in which the copper or aluminium bars are placed?**

Ans. Just as the cage of a squirrel and the rotor construction is known as squirrel cage rotor.

Q. **What do you understand by the wound rotor?**

Ans. If in the slots of a rotor, the winding is placed and brought to sliprings, that rotor is called wound rotor.

Q. **What types of slots are used in the rotor of slipring induction motor?**

Ans. Open type slots.

Q. **Why the rotor is made of laminations?**

Ans. To reduce the eddy current losses.

Q. **What is done to reduce the hysteresis losses?**

Ans. Some percentage of silicone is added.

Q. **Which material is used for short circuited bars?**

Ans. Copper or aluminium bars.

Q. **Sometimes it is observed that the bars are skewed, why?**

Ans. To avoid magnetic interlocking between the stator and rotor and to increase the length of the rotor conductor to enable the production of more torque.

Q. **What types of slots are used for stator of medium size of squirrel cage motor?**

Ans. Semi-open type.

Q. **Name the slot insulation generally used for medium size squirrel cage motor?**

Ans. Leatheroid or milliax paper.

Q. **How many terminals are brought out from the slipring induction motor?**

Ans. Three from stator and three from rotor and sometimes six from the stator winding and three from the rotor winding.

Q. **How many terminals are brought out from the three phase squirrel cage motor?**

Ans. Three or six.

Q. **If a three phase induction motor is to be started from (a) D.O.L. starter, (b) star delta starter, how many terminals are to be brought out?**

Ans. (a) Three.
(b) Six.

Q. **Name the type of bearings generally used for medium size of squirrel cage induction motor?**

Ans. Ball-bearings.

OPERATION

Q. **What is the nature of the magnetic field produced because of the three phase stator winding?**

Ans. Rotating magnetic field.

Q. **What are the characteristics of the rotating magnetic field?**

Ans. It has constant magnitude rotating at a constant speed, i.e. synchronous speed.

Q. **Can you change the d.o.r. of the rotating magnetic field?**

Ans. Yes, by interchanging the two phases (line).

Q. **What will be the direction of rotation of the rotor with respect to the rotating magnetic field?**

Ans. Same as that of rotating magnetic field.

Q. **If you change the d.o.r. of the rotating magnetic field, what will happen to the d.o.r. of the rotor?**

Ans. It will also change.

Q. **How many electrical degrees are there in a four pole machine?**

Ans. 720° elect.

Q. **How the electrical degrees are calculated?**

Ans. 360 × No. of pair of poles.

Q. **Define synchronous speed?**

Ans. The speed of the rotating magnetic field is known as the synchronous speed. It is represented by N_s and measured in r.p.m. It is given by the formula

$$N_S = \frac{120f}{P} \text{ rpm.}$$

Q. **On what factors does the synchronous speed N_S depend?**

Ans. Frequency and number of poles.

Q. **If you increase the number of poles of a motor will the speed decrease or increase?**

Ans. Speed is inversely proportional to the number of the poles so the speed will decrease with the increasing of the poles.

Q. **What is the angular displacement between the magnetic field produced by each phase?**

Ans. 120° elect.

Q. **What is torque?**

Ans It is defined as the turning or twisting moment of a force about its axis.

Q. **In induction motor, on what factors does the torque depend?**

Ans Flux produced by the stator winding, the rotor current and rotor power factor.

Q. **Can the rotor speed by equal to synchronous speed in case of induction motors?**

Ans. No.

Q. **Why the N_r (rotor speed) and N_S (synchronous speed) cannot be equal?**

Ans. Because in induction motor, the induction can only be possible if there is any relative motion between the rotor bars and the magnetic field so if there is no induction, there will not be any e.m.f. induced in rotor bars, there will not be any current and hence no torque. So N_r cannot be equal to N_s.

Q. What do you know about the slip speed?

Ans. It is the difference between the rotor speed and synchronous speed.

Q. Can you tell the formula for calculating the percentage slip of the motor?

Ans. $\% \, s = \frac{N_s - N_r}{N_s} \times 100.$

Q. Has the e.m.f. induced in the rotor winding any frequency?

Ans. Yes.

Q. What is the rotor frequency and on what factors does it depend?

Ans. Rotor frequency = Slip × Supply frequency. It depends upon the slip and supply frequency both.

Q. What is the rotor frequency at the starting of a 50 c/s squirrel cage induction motor?

Ans. It is 50 c/s but decreases with the increase of the speed of the motor.

Q. Why in running condition the rotor frequency is less?

Ans. Because in running condition the slip is less and thus the flux linking the rotor conductors has less speed producing less frequency.

Q. What do you mean by the fractional slip?

Ans. If the slip speed is shown in terms of synchronous speed as

$$= \frac{N_s - N_r}{N_s}$$

it is called as fractional slip.

Q. Will there be any induced emf in rotor winding or an induction motor, if its rotor is rotated at synchronous speed?

Ans. No, there will not be any emf induced.

Q. Why will there be no induced emf?

Ans. Because the speed of the magnetic field produced is the synchronous speed and the rotor speed is also same so there will not be any linkage and thus no emf will be induced in the rotor.

Q. In terms of percentage slip language what do you call this situation?

Ans. Zero slip situation.

Q. How the power factor of the rotor winding is calculated?

Ans. $\text{Rotor power factor} = \frac{\text{Rotor resistance}}{\text{Rotor impedance}}$

Q. Do you know something about sX_2?

Ans. It is the slip multiplied by the rotor reactance. It is the reactance of the rotor in running condition.

Q. Give any formula for calculating the torque in an induction motor?

Ans. $T \propto I_r \cos \phi_r$

or $$T = \frac{\text{Rotor input}}{2\pi N_s}$$

or $$T = \frac{\text{Rotor output}}{2\pi N_r}$$

Q. **How the starting torque of an induction motor is improved?**

Ans. By increasing the resistance of the rotor winding at the time of starting (slipring induction motor).

Q. **How slip is concerned with the torque?**

Ans. Torque $\propto$ Slip.

Q. **In case of squirrel cage motor if the resistance of rotor bars is increased how will it affect the performance of the motor?**

Ans. Though by increasing the rotor resistance, the torque developed in starting is increased, but the efficiency of the motor will decrease because of losses in running conditions.

Q. **What is the effect of change in voltage on the production of torque of an induction motor?**

Ans. Torque is directly proportional to the square of voltage.

LOSSES

Q. **What are the variable losses and why these are called so?**

Ans. Generally, the copper losses are called the variable losses as these losses are proportional to the square of current.

Q. **Which losses are constant losses and why?**

Ans. The stray losses, frictional losses and windage losses are the constant losses. These are proportional to the speed of the motor which is almost constant hence these are called constant losses.

Q. **What is efficiency?**

Ans. It is the ratio of output to input generally shown in percentage.

Q. **Can you calculate the output of a machine if the losses and input are given?**

Ans. Yes. Output = Input – Losses.

CHARACTERISTICS

Q. **What are the characteristics of an induction motor?**

Ans. The behaviour and performance of the induction motor at different load currents and speed are termed as characteristics. These are :

1. Speed and load current characteristic
2. Torque and load current characteristic

3. Speed and torque characteristic.

Q. **What is the speed load characteristic?**

Ans. It shows the relationship between the speed and the load put on the motor.

Q. **What is torque and speed characteristic?**

Ans. It shows the relationship, how much load is put on the machine and what will be the speed of the motor on that load.

Q. **What is the basic differences between single cage and double squirrel cage induction motors?**

Ans. In double squirrel cage rotor, there are two cages one of high resistance other of low resistance. In case of single cage rotor, it has only one cage of low resistance winding.

Q. **What is the utility of providing double cages in the rotor?**

Ans. The motor has better starting torque.

Q. **How is the running efficiency of single cage and double cage rotor motor?**

Ans. Double cage rotor motor has low efficiency than the single cage rotor motor.

Q. **Name the induction motor which has better starting and good efficiency?**

Ans. Slipring induction motor.

Q. **Why the slipring induction motor has better starting torque and good efficiency?**

Ans. Because at the time of starting the external resistance is introduced in the rotor circuit which causes better starting torque and after that the external resistance is removed and motor works as the squirrel cage induction motor.

STARTERS

Q. **What is a starter?**

Ans. The starter is a safety device which is used for the starting of motor.

Q. **Name the protective devices used in the starter?**

Ans. NVRC, OLRC, Push button station etc.

Q. **What are the NVRC and OLRC?**

Ans. These are no volt release coil and overload release coils; the no volt release coil protects against supply failure and overload release against the overloading.

Q. **What are the different types of OLRC?**

Ans. 1. Thermal or bimetallic relay.

2. Magnetic relay.
3. Desh pot type magnetic relays.

Q. **What do you understand by the thermal or bimetallic relay?**

Ans. It is made by two different metals. It has two strips of different temperature coefficient substances. As the temperature increases the strip bends towards the material having less temperature coefficient. This principle is used in bimetallic or thermal relays.

Q. **What is magnetic relay and how it protects motor?**

Ans. It works on the principle of magnetism. Whenever normal current is flowing, the coil will produce a magnetic flux but it will not attract the arm attached. Whenever current exceeds the limit, the flux produced will be certainly more and will attract the arm which will ultimately trip the contractor and protect the motor against overloading.

Q. **On what principle the desh pot type relay works?**

Ans. Magnetic effect of current.

Q. **What is the unique characteristic of the desh pot type relay?**

Ans. It is having an arrangement of time lag.

Q. **How the time lag is adjusted?**

Ans. By adjusting either the holes of different diameters attached with moving or attracting system or by adjusting the distance of the soft iron rod and coil.

Q. **What is the function of no volt releases coil? What is the resistance of this coil?**

Ans. The no volt release coil protects the motor from supply failure. It has more resistance. It is wound with thin conductor having more number of turns and it is connected across the line.

Q. **What type of starters are used for starting the three phase motors?**

Ans.
1. Direct on line starter (DOL starter).
2. Star-delta starter.
3. Autotransformer starter.
4. Slipring induction motor starter.

Q. **Which type of starters are preferred for 1 HP, 3 HP and 10 HP motors?**

Ans. One HP motor can even be started without starter but should be started with DOL starter. 3 HP with DOL starter, 10 HP with star-delta starter.

Q. **How many types of contacts are there in a DOL starter?**

Ans. Fixed contacts, moving contacts and auxiliary contacts.

Q. **What is the function of auxiliary contact.**

Ans. In case when the starter is on, the auxiliary contacts will be in parallel with the start button and the supply will be continued to the no volt release coil through these contacts.

Q. **What is the remote controlling?**

Ans. To control the motor from more than one place is known as remote controlling?

Q. **How all the start and stop buttons are connected in remote controlling?**

Ans. All start buttons are connected in parallel and stop buttons in series.

Q. **What is the phase sequence?**

Ans. The systematic arrangement of connections of all phases is known as phase sequence. Let us say in three phase winding if phase I is starting from $\theta°$ elect., phase II will start from $(\theta + 120)°$ elect. and phase III from $(\theta + 240)°$ elect. This arrangement of winding is known as phase sequence.

Q. **What will happen if the phases are not arranged in sequence?**

Ans. Motor will not run properly and operations will not be smooth. The motor will make abnormal sound and may blow fuses even.

Q. **Why the D.O.L. starter is called so?**

Ans. In this starter the same supply voltage is fed to the motor, that is why it is called as direct on line starter.

Q. **What is a star delta starter?**

Ans. It is the single starter in which the motor is connected in star at the time of starting and in delta in running condition. These can be manual or automatic.

Q. **When the motor should be put in delta after starting in star?**

Ans. As soon as the motor attains 75% of the synchronous speed the motor should by connected in delta.

Q. **If the handle is moved slowly to the delta position is it possible and advisable?**

Ans. No, in the star delta starter it is not possible, there a mechanical lever protection is given, so that it should not be delayed. If it is moved slowly the lever will not allow to connect the motor in delta.

Secondly if the speed drops down because of time lag, the current will increase more than the predetermined value if it is now connected in delta. So that time limit is there, one should be quick enough to connect the motor into delta.

Q. **How the changeover is governed in case of automatic star delta starter?**

Ans. A timer is provided which governs the changeover time of delta.

Q. **What type of timers are used?**

Ans. Generally motorised or bimetallic.

Q. **Why it is essential to start the motor more than 5 HP with star-delta starter?**

Ans. To limit the starting current.

Q. **How the star-delta starter reduces the starting current?**

Ans. By connecting the motor in star at the time of starting, the voltage per phase is reduced, so motor will take less current than that if it would have been started in delta.

Q. **How many times it reduces the starting current than if it would have been connected in delta?**

Ans. About three times.

Q. **What is an autotransformer starter?**

Ans. The starter in which autotransformers are used to reduce the impressed voltage is known as autotransformer starter.

Q. **How it differs from the star-delta starter?**

Ans. In case of star-delta starter, the starting voltage to phase winding is only $V_1/\sqrt{3}$ but in autotransformer depending upon the requirement the voltage can be adjusted. In running condition both are giving full voltage to the motor.

Q. **In which connections the autotransformer windings are connected?**

Ans. In star.

Q. **Can you tell whether the transformer remains energised throughout the operation of the motor or not?**

Ans. No, only at the time of starting and de-energised during running position.

Q. **The no-volt release coil remains energised throughout the operation or not?**

Ans. Yes, remains energised.

Q. **Can merely a D.O.L. starter be used for starting a slipring induction motor?**

Ans. No, for that a slipring induction motor starter is used.

Q. **What precautions are to be taken while starting a slipring induction motor?**

Ans. The complete rotor resistance must be in rotor circuit and then only put the starter in on position.

Q. **What are the general parts of slipring induction motor starter?**

Ans. The parts are rotor resistance which is variable, brass studs, handle. OLRC, NVRC, overload tripping contact, two extra push buttons, stop button, metallic container, and connecting terminals.

Q. **What is the function of start push button?**

Ans. The motor will not start unless the whole rotor resistance is not in rotor circuit of slipring induction motor.

Q. **What is the function of second push button, generally which is being used (pressed) while moving the handle of variable resistance ahead?**

Ans. It is connected in series with no volt release coil and when pressed the coil of starter (DOL starter) remains energised otherwise the movement the handle is pushed forward and start button which is a NO contact will open and deenergise the no volt release coil.

Q. **Which types of overload release coil are used in this starter?**

Ans. Nowadays where DOL starter plunger is used for main line, thermal type overload relays are used. But in old starter magnetic overload relays were used.

Q. **Initially if the starter handle is in between ON and OFF position and you press start button, will the motor start?**

Ans. No, the NVRC circuit is open-circuited and the plunger will not operate.

Q. **Can you change the speed of the motor if so desired?**

Ans. In case of slipring induction motor the speed can be regulated up to some extent say 5%.

Q. **What is the utility of this starter?**

Ans. It reduces the starting current with high starting torque.

Q. **Why the starting torque is increased?**

Ans. Because the starting torque is directly proportional to the rotor circuit resistance. The rotor resistance is increased by increasing additional resistance in the rotor circuit.

Q. **Can't you increase the rotor winding resistance permanently or use more resistance short circuiting bars?**

Ans. No, it is not advisable because at the time of starting the torque will be more but in running condition the rotor copper losses will be considerably high which will decrease the efficiency.

Q. **How will you calculate copper losses of this machine?**

Ans. The rotor circuit current I_R, rotor circuit resistance $(R_r + R_{ex})$ so the starting losses are $I_R^2(R_r + R_{ex})$ and running condition current $= I_R^2 \cdot R_r$ watts.

SPEED CONTROL

Q. **What are the methods of speed control?**

Ans. Pole changing, frequency changing, rheostatic control and cascade control.

Q. **If the frequency is increased will the speed increase?**

Ans. Yes, the speed is directly proportional to the frequency.

Q. **How the speed is linked with the poles?**

Ans. By increasing the number of poles the speed will decrease and vice versa.

Q. **What do you mean by rheostatic control?**

Ans. The speed is controlled by introducing a rheostat either in rotor circuit or in stator circuit. This way the speed be controlled below normal.

Q. **Why the rheostatic control is not commonly used?**

Ans. Because of more copper losses (I^2R).

Q. **What is the condition for maximum efficiency?**

Ans. The copper losses should be equal to the iron losses.

Q. **What is reversing switch?**

Ans. The switch which changes the d.o.r. of the motor by putting it into the other position is known as reversing switch.

Q. **How the d.o.r. is changed by reversing switch?**

Ans. By putting it into the other position the phases are interchanged and thus the d.o.r. is changed.

Q. **What is the speed regulation?**

Ans. It is the ratio of the increase in speed when the full load is thrown off, shown in the percentage of the no load speed.

i.e. $\% \text{ Regulation} = \dfrac{N_0 - N_L}{N_0} \times 100$

Q. **How the speed of an induction motor is measured?**

Ans. By means of speedometer or tachometer.

Q. **What are the methods of finding out the slip?**

Ans.
1. By means of measuring rotor speed.
2. By stroscopic effect.

Q. **How the slip is calculated by means of the tachometer?**

Ans. Measure the rotor speeds and as we known the poles that means the synchronous speed of the motor can be calculated so substitute in the formula $\dfrac{N_s - N_r}{N_s} \times 100$ and calculate the slip.

Q. **In star-delta starter, how the no volt release coil is connected?**

Ans. It is connected across the phases i.e. line in delta condition and across phase voltage in starting condition.

Q. **Where the overload release contact is located?**

Ans. Generally on left hand side in Striplex and Bantex starters. But it is always connected in series with no volt release coil.

Q. **How overloading is protected?**

Ans. In case of overloading, the thermal overload relays carry more current.

As a result more heat is developed causing the bending stresses. The bending will operate the overload releasing bar i.e. contact to open and deenergising the no volt release coil, as a result the handle will come to off position and motor will be protected.

Q. **Can you adjust the overload current?**

Ans. Yes, it can be easily adjusted according to the overloading capacity or HP of motor.

Q. **The overload tripping contact bar is a straight rod with three lever type overload relays, can you tell the tripping contact will only open if all the three overload relays operate?**

Ans. No, the bar is placed in such a way that it will open if either of the three relays operate. It is a protection, and each individual overload relay is connected in series of each phase. So if overloading is caused in either of the phases it will operate and deenergise NVRC.

Q. **If the terminals of the starter are not known how will you find out?**

Ans. In this starter there are nine terminals brought out, three for lines and six terminals for motor windings. Initially put the handle in star and see with the series testing lamp the there terminals which are short circuited by thick strip i.e. star connections.

Now when three ends are known whether starting point or end points, the remains could be easily traced out. Let the first terminal is marked as A_1 then second is B_1 and third is C_1. In delta connections A_1 is short circuited with B_2 or C_2. Let it be B_2, then B_1 will be connected with C_2 and C_1 with A_2. Now check which terminal is short circuited by the same strip and mark it B_2. Similarly trace C_2 and A_2. Now all L_1, L_2 and L_3, A_1, B_1, C_1, A_2, B_2, C_2 are known. Connect the motor and start.

Q. **In automatic star-delta starter this facility of handle is not there, so how will you identify the terminals?**

Ans. In that case, a contactor, generally the last one, is of star connections, so identify the terminals and connected leads. Then see the second contactor and note the terminals which are to be short circuited. When star contactor opens, they will be dissimilar ends. Remember that A_1 is short circuited with B_2, B_1 with C_2 and C_1 with A_2 or A_1 is with C_2, C_1 with B_2 and B_1 with A_2. Identify the terminals and connect with respective terminal.

Q. **Which part of the starter helps for connecting the automatic star delta starter in delta after starting?**

Ans. Timer. The time can be adjusted after how many seconds of starting the motor is to be connected in delta.

Q. **A squirrel cage induction motor is made to have more air gap clearance. Will it draw more current or less current than the determined value?**

Ans. It will draw more current and the windings will be heated up.

Q. **A DOL starter while switched on and start button is pressed, the plunger is energised but motor is not running what could be the possible fault?**

Ans. The outgoing leads are disconnected or broken or damaged. The contacts are not coming in contact. The short circuiting strips of plunger are damaged or broken. The input is not coming or open circuited. Either of these or in conjunction may be the reason.

Q. **While the start button is pressed the motor gets accelerated but stops as soon as the button is released?**

Ans. The auxiliary contact of the starter is not working satisfactory. As this is in parallel with the start button so as long as the start button is on, supply continues to no volt release coil but breaks with its releasing because the supply of NVRC is not continued so the plunger is released and stops the motor.

Q. **Generally for a 5 HP induction motor, which type of slots are used?**

Ans. Semiclosed slots with parallel teeth.

Q. **What are the types of slots used for the starter and rotor windings of the induction motor?**

Ans. Closed type, open type and semiclosed types.

Q. **How the starting torque of an induction motor can be increased?**

Ans. Starting torque is directly proportional to the square of voltage and rotor resistance. Generally rotor resistance is increased (slipring induction motor application).

Q. **Can you roughly tell, at what load the power factor of the induction motor will be maximum.**

Ans. Full load capacity as these are designed for that.

Q. **Can two A.C. induction motors of same h.p. having different efficiencies be started from same types of starters?**

Ans. Yes, they can be started as the efficiencies are different not the capacity of the motors.

Q. **What is the rotor impedance of an induction motor in standstill condition?**

Ans. It is $Z_r = \sqrt{R_r^2 + (sX_r)^2}$

At standstill condition $s = 1$

So $Z_r = \sqrt{R_r^2 + X_r^2}\ \Omega$.

where Z_r = rotor impedance in ohm, R_r = rotor resistance in ohm, and X_r = rotor reactance in ohm.

What is the value of Z_r in running condition?

Ans. $Z_r = \sqrt{R_r^2 + (sX_r)^2}$

where s is the slip of the motor at that speed.

Q. **What is the expression for the rotor power factor in case of motor running with S-slip?**

Ans. $\cos \phi_r = \dfrac{R_r}{\sqrt{R_r^2 + (sX_r)^2}}$.

Q. **Which change R_r, s, X_r will change/affect the maximum on the rotor power factor?**

Ans. Change in rotor resistance because R_r is directly proportional to the power factor.

Q. **The rotor resistance is also in the denominator why don't you consider the net effect will be neutralized?**

Ans. No, the net effect will be a corresponding change, because the rotor resistance square under-root will decrease the denominator more than the numerator.

FAULTS

Q. **What are the possible faults in an induction motor?**

Ans. Earth fault, short circuit between the coils, open circuit in the coil, leakage or failure of the insulation etc.

Q. **How earth fault can be tested?**

Ans. By means of test lamp or by means of megger. Connect one terminal on the body the other on the winding. In case of series lamp testing, the lamp will glow and in case of megger testing the megger will give zero reading.

In case of lamp testing method one thing should be kept in mind that the phase wire may not be directly earthed, phase must be connected through lamp.

Q. **How much leakage current is allowed?**

Ans. 1/5000 of the total full load current of the appliance/motor.

Q. **What are the possible tests to be carried out for the proper functioning of the motor?**

Ans.
1. Insulation resistance test.
2. Phase sequence test.
3. The leakage current should be under permissible limits.

Q. **Why the windings are varnished?**

Ans. To avoid air bubbles in the coil turns, to give sufficient mechanical strength to the conductors and to avoid moisture also.

Q. **Why the motor winding and varnish are heated up before varnishing?**

Ans. So that even the small bubble of air from the coil may come out and the hot varnish may penetrate to the deep and grip the conductors in effective manner. Thus the winding will be mechanically strong enough.

Q. **Why the no load test is conducted?**

Ans. To find out the no load current I_0, the no load power factor cos ϕ, windage and friction losses, no load copper losses, no load input to the motor, no load resistance R_0 and reactance X_0.

Q. **What for short circuit test is performed for squirrel cage induction motor?**

Ans. Just to find out short circuit current with normal applied voltage, power factor on short circuit, total equivalent resistance and reactance of the rotor referred to stator winding.

Q. **How the short circuit test is conducted?**

Ans. The rotor is held firmly and stator is connected across variable voltage. It is just equivalent to the short circuit test of the transformer.

SINGLE PHASE INDUCTION MOTORS

Q. **What is single phase induction motor?**

Ans. The motor which works on single phase supply and converts it into mechanical energy is known as single phase motor.

Q. **What are the classifications of single phase motors?**

Ans.
1. Synchronous motor.
2. Induction motor.
3. Commutator motor.

Q. **What do you know about the commutator motor?**

Ans. The motor in which rotor is wound and the leads are brought to the commutator is known as commutator motor. For example, series motor and repulsion motor.

Q. **How the repulsion motor works?**

Ans. It works on the principle of repulsion between the magnetic field produced by the stator and rotor windings.

Q. **Name the different classifications of the repulsion motor?**

Ans.
1. Plane repulsion motor.
2. Repulsion induction motor.
3. Repulsion start induction run motor.

Q. **When the torque is developed for rotation in repulsion motor : (a) the brushes are at MNP; (b) the brushes are at magnetic axis; (c) in between the MNP and magnetic axis?**

Ans. In between the magnetic axis and magnetic neutral axis.

Q. **What is the nature of the magnetic field produced by the single phase alternating current?**

Ans. Alternating magnetic field.

Q. **Is this magnetic field a rotating magnetic field?**

Ans. No.

Q. **Are the single phase motors self-starting?**

Ans. No, these are not self-starting.

Q. **How you are getting the splitting of the phases?**

Ans. By providing two windings known as the starting winding and running winding.

Q. **What are the methods of obtaining starting torque in split phase motors?**

Ans. The resistance start, the reactance start, the capacitor motors (capacitor start capacitor run, capacitor start induction run and permanent capacitor motor).

Q. **Can you give an example of permanent capacitor motor?**

Ans. Ceiling fan or table fan.

Q. **Where the capacitor start capacitor run motors are used?**

Ans. Places where high starting torque is required for example compressors etc.

Q. **Generally which single phase motor is used in lathe machines?**

Ans. Capacitor start induction run.

Q. **In case of capacitor start induction run motor the starting winding is cut off in running condition. How is it possible?**

Ans. It is possible by providing a switch generally known as centrifugal switch. In normal condition it is closed and in operation it is opened.

Q. **How will you identify the starting winding and running winding?**

Ans. The starting winding has high resistance and the test lamp will give dim light and the running winding has less resistance the test lamp will give bright light.

Q. **Where the start winding in case of shaded pole motor is placed?**

Ans. In the groove, where the pole faces are shaded.

Q. **How much pole face is shaded?**

Ans. It is 1/4th or 1/3rd of the pole face.

Q. **The poles are projected or smooth?**

Ans. Projected.

Q. **Which action of the magnetic field is responsible for the production of torque in shaded pole motors?**

Ans. Sweeping action of the magnetic field.

Q. **The shading ring is made of which material?**
Ans. Generally copper.

Q. **Where single phase synchronous motors are used?**
Ans. Generally for constant speed for example in clocks etc.

Q. **In which winding circuit the capacitor is used and why?**
Ans. The capacitor is used in series with the starting winding to give a phase displacement to the current drawn by starting and running windings.

Q. **In a capacitor start motor, if the staring winding is not disconnected what will happen?**
Ans. The starting winding circuit is for starting purpose. It has to be disconnected as soon as motor attains starting. If it is not disconnected the motor will give abnormal sound and will take more current and finally may burnt out.

Q. **Is there any such single phase capacitor motor in which the starting winding circuit is not disconnecting even in normal working?**
Ans. Yes, there are, for example a ceiling fan, a exhaust fan etc.

Q. **Why the fault of abnormal sound and taking more current is not experienced?**
Ans. It is because the running torque of these fans are calculated on that assumptions that the flux will be produced by both the windings in normal working condition. The designing is such that both windings will remain in the circuit during starting and running conditions.

Q. **Sometimes we have seen that a ceiling fan suddenly starts moving in opposite direction with slow speed what could be the defect?**
Ans. The starting winding circuit is open-circuited may be due to burnt winding or damaged capacitor or the running winding is damaged. The fan is running only on one winding. The angle of blades and thrust of air in the room accelerated the blades in opposite direction. As we know that the single phase motors are not self-starting, they will continue to move in the direction in which these are initially accelerated.

Q. **Why a table fan while switched on sticks to one side and does not work (revolve)?**
Ans. The brushes are defective and need replacement (the other things oscillating attachment being alright).

Q. **The torque developed in A.C. single phase motors is proportional to the sine or cosine of the angle between starting winding current and running winding currents?**
Ans. Sine of the angle between starting winding and running winding currents.

Q. **Generally which type of capacitor is used as a starting capacitor in A.C. single phase motors?**
Ans. Electrolytic capacitor.

Q. **Can an induction motor be used for a record player or tape recorder?**

Ans. No, generally single phase hysteresis motor is used for this purpose.

Q. **Can you specify a particular use of reluctance motor?**

Ans. It is a suitable motor for signalling device purpose.

Q. **Which motor is suitable for portable tools, the A.C. split phase induction motor, universal motor or shaded pole motor?**

Ans. Universal motor.

Q. **The starting torque of single phase induction motor is how many times the full load torque?**

Ans. The starting torque of A.C. single phase induction motor depends upon the type of starting.

Q. **For changing the d.o.r. of a split phase induction motor what will you do?**

Ans. The DOR of a split phase induction motor can be changed by changing the direction of current in either of starting or running winding. Generally the starting winding direction is reversed.

Q. **Which capacitor is used in shaded pole motor?**

Ans. Capacitor is not used in shaded pole motor.

Q. **How many sets of wound coils have a four pole shaded pole induction motor?**

Ans. Four sets of wound coils.

Q. **The main purpose of shaded pole winding is ......?**

Ans. To provide starting winding purpose and to facilitate the starting torque.

Q. **The capacitor helps in splitting single phase supply into two phases for producing rotating magnetic field?**

Ans. Yes, it helps and splits for the production of torque.

Q. **Can a universal motor be started by 3-phase supply?**

Ans. No, the universal means (here) that this motor can work on A.C. single phase or D.C. both.

SYNCHRONOUS MOTOR

Q. **What do you understand by a synchronous motor?**

Ans. The motor which rotates at the synchronous speed is known as synchronous motor.

Q. **Can you tell, whether the synchronous motor is a self-starting one?**

Ans. No, it is not self-starting.

Q. **Can you name a single motor in three phase motors where two supplies work simultaneously?**

Ans. Yes, it is synchronous motor.

Q. **On which principle does the synchronous motor work?**
Ans. Magnetic interlocking.

Q. **What are the types of synchronous motor?**
Ans.
1. Rotating magnetic field type.
2. Stationary magnetic field type.

Q. **What are the starting methods of the synchronous motor?**
Ans.
1. By means of a separate motor.
2. By means of an induction motor action.
3. By means of a pony motor.

Q. **What are the dampers or damping winding?**
Ans. It is a winding placed in the pole faces and short-circuited at both the ends as that of a squirrel cage winding. It is used for starting purpose and to damp or hunt the oscillations which are caused by the load impacts or removal of loads.

Q. **Do you know the term 'angle of coupling'?**
Ans. It is the maximum angular displacement between the magnetic axis of stator and rotor poles with which a synchronous motor can be loaded without losing the synchronism.

Q. **What is pull in torque?**
Ans. The amount of torque that a motor is able to exert when pulling into synchronism is called the pull in torque.

Q. **What is pull out torque?**
Ans. The maximum torque that a motor can develop without losing the synchronism is known as pull out torque.

Q. **If you increase the load on the synchronous motor, what will happen, will the speed drops down or not?**
Ans. The speed will remain the same synchronous speed but the angle of coupling will increase.

Q. **What will be the current at unity power factor?**
Ans. It will be lowest possible.

Q. **If there is a constant load on the motor and you are changing the field excitation, what will happen to the load current?**
Ans. The current will be more when the motor is over or under excited, but speed will be same.

Q. **Will the overexcited synchronous motor improve the power factor?**
Ans. Yes, in that case it is known as synchronous capacitor.

Q. **Where the synchronous motor is used?**
Ans.
1. To improve the power factor.
2. To drive the load.
3. For constant speed.

Q. **If the current drawn is plotted under the overexcited and under-excited conditions, the curve so obtained will have which shape and what is it called?**

Ans. V-shape and the curve is known as V-curve of the synchronous motor.

Q. **What are the advantages of this motor?**

Ans.
1. Constant speed.
2. Power factor correction.

Q. **What are the disadvantages of this motor?**

Ans.
1. Speed cannot be controlled.
2. More maintenance is required.
3. It is not a self-starting motor so starting means is required.
4. Cannot be started with load.

Q. **Generally which starter is used for starting the synchronous motor?**

Ans. Autotransformer starter.

Q. **Why oscillations are produced in synchronous motor?**

Ans. Because of variation of load.

Q. **How hunting is achieved?**

Ans. By using damping winding in the pole faces.

Q. **If you change the field excitation how much variation in speed you can get?**

Ans. No variation of speed.

Q. **While changing the field excitation, which factor is effected?**

Ans. Power factor.

Q. **Which motor may have high efficiency, the synchronous motor or induction motor?**

Ans. Synchronous motor.

Q. **If the field windings are connected in reverse order will the motor start and if so, will it produce its full load torque?**

Ans. The motor may start, but will not produce its full torque.

OBJECTIVE TYPE MULTIPLE CHOICE QUESTIONS

1. The rotor of A.C. induction motor is not connected to ......
   (a) D.C. supply (b) any supply
   (c) 50 c/s A.C. mains (d) A.C. 50 c/s 400 V.
2. The squirrel cage induction motor has ......
   (a) small starting torque (b) high starting torque
   (c) extra means for starting (d) no starting torque

3. The speed of rotating magnetic field of A.C. in polyphase induction motor can be calculated by using ...... formula

   (a) $N_s = \frac{120P}{f}$ (b) $N_s = \frac{120f}{P}$

   (c) $N_s = 120P \times f$ (d) $N_s = \frac{1}{120f \times P}$

4. The skewing in the rotor or stator of an induction motor is given ......
   (a) to help in reducing the interlocking tendency
   (b) to increase the speed of the motor
   (c) to reduce the speed
   (d) to reduce magnetic humming and interlocking tendency.

5. In a three phase induction motor the resultant magnetic field is of constant magnitude i.e.
   (a) equal to maximum flux, ϕ_m
   (b) 1.5 times the maximum value of flux due to any one of the phases
   (c) three times the flux produced by a single phase
   (d) 0.866 times the maximum flux.

6. The e.m.f. induced in the rotor winding of the A.C. motor has the frequency equal to ......
   (a) slip × supply frequency
   (b) slip × maximum flux produced
   (c) $\frac{1}{\text{slip}}$ × supply frequency
   (d) ϕ_m × supply frequency.

7. The synchronous speed of a 4 pole 50 c/s 400 V 10 HP motor is ......
   (a) 750 r.p.m. (b) 1500 r.p.m.
   (c) 600 r.p.m. (d) 3000 r.p.m.

8. The rotor speed of a 4 pole A.C. induction motor can be ......
   (a) 766 r.p.m. (b) 1440 r.p.m.
   (c) 1510 r.p.m. (d) 1550 r.p.m.

9. The torque and voltage has a relation in A.C. induction motor ......

   (a) $T \propto V^2$ (b) $T \propto \sqrt{V}$ (c) $T \propto V^{1.5}$ (d) $T \propto \frac{1}{V^2}$

10. The slipring induction motor can be started by ......
    (a) D.O.L. starter (b) star delta starter
    (c) autotransformer starter (d) rotor resistance type starter.

11. The speed of a squirrel cage induction motor can be changed by ......
    (a) rheostatic control

(b) frequency control
(c) changing the number of conductors in stator winding
(d) changing the number of parallel paths.

12. The synchronous motor has ......
(a) constant speed
(b) variable speed
(c) constant poles
(d) constant size.

13. The overexcited synchronous motor ...... the power factor of the system.
(a) maintains the same
(b) decreases
(c) improves
(d) None of the above a, b, c.

14. The torque of squirrel cage induction motor under running condition is maximum when the rotor phase values are ......
(a) $X_2 = R_2$
(b) $X_2 > R_2$
(c) $X_2 < R_2$
(d) Not performing any contribution.

15. The maximum torque in running condition of a three phase squirrel cage induction motor ......
(a) is directly proportional to R_2
(b) is indirectly proportional to R_2
(c) is independent of R_2
(d) is inversely proportional to the X_2.

16. In case of squirrel cage induction motor, if the rotor e.m.f. frequency is equal to the supply frequency, then the rotor is ......
(a) standstill
(b) running at N_s
(c) running below N_s
(d) running above N_s.

17. In case of double squirrel cage rotor motor, the inner cage of the motor has ......
(a) high resistance low inductance
(b) low resistance low inductance
(c) high resistance high inductance
(d) low resistance high inductance.

18. The ratio of actual speed to synchronous speed in case of an induction motor is also called ......
(a) slip speed
(b) slip
(c) motor efficiency
(d) rotor efficiency.

19. The three phase squirrel cage induction motor takes approximately its full load current ...... times the h.p. of the motor.
(a) double
(b) one and half
(c) three
(d) six.

20. A one H.P. three phase squirrel cage induction motor takes approximately 1.6 amp. full load current, it should be tested by a megger of at least ...... volts.

    (a) 500 V (b) 1000 V
    (c) 2000 V (d) None of the above.

21. In an induction motor if the number of rotor slots are equal to the number of stator slots the motor will ......

    (a) run smoothly (b) run with 50% slip
    (c) may refuse to start (d) run over the N_S.

22. In case of induction motor, skewing of rotor slot actually helps in ...... of motor.

    (a) starting (b) quick stopping
    (c) suppressing undesired harmonics
    (d) reducing noise.

23. Which motor is generally used for electric shavers?

    (a) Shaded pole (b) Universal motor
    (c) D.C. motor (d) None of the above.

24. The disadvantages of shaded pole motor are ......

    (a) low starting torque (b) low efficiency
    (c) low p.f. and overload capacity
    (d) All of the above.

25. Shaded pole motors are provided with ......

    (a) capacitor (b) centrifugal switch
    (c) squirrel cage rotor (d) commutator.

CHAPTER 19

Electrical Measuring Instruments

INTRODUCTION

Q. **What are electrical measuring instruments?**

Ans. The instruments which are used to measure the electrical quantities, viz current, voltage, power, frequency, power factor etc. are known as electrical measuring instruments.

Q. **Does it actually measure or compare the quantities?**

Ans. Compare.

CLASSIFICATION

Q. **What are the absolute instruments?**

Ans. The instruments which give the quantity to be measured in terms of instrument constant and its deflection. Such instruments do not require any comparison with any standard instruments.

These are rarely used and kept in laboratories for standardisation of the other instruments.

Q. **What are secondary instruments?**

Ans. The instrument in which the deflection gives the magnitude of the electrical quantity to be measured directly.

Q. **How the instruments are classified?**

Ans. According to the type, working principle and accuracy.

Q. **What are the classifications according to the accuracy?**

Ans.
1. Standard instruments.
2. Sub-standard instruments.
3. Commercial instruments.

Q. **What is the percentage of error possible in the standard, sub-standard and commercial instruments?**

Ans. 0%, about 0% and ±3%, respectively.

Q. **Which instruments are used to calibrate the commercial instruments?**

Ans. Sub-standard instruments.

Q. **Why the standard instruments are made?**

Ans. To have a standard and are kept in research laboratory for standardizing the sub-standard instruments.

Q. **Can you use standard instrument for commercial purpose?**

Ans. No.

Q. **What are the instruments according to the function?**

Ans. Indicating type, integrating type and recording type.

Q. **What do you understand by the indicating type instrument?**

Ans. The instrument which indicates the reading by means of the deflection of a pointer and can be read on the calibrated scale, is known as indicating type instrument.

Q. **What are the recording type instruments?**

Ans. The instrument in which the readings are recorded with the help of a pointer (pen) and on a graph. These instruments can indicate the particular reading at a particular time.

Q. **Name the type of the instrument which integrates the reading or measures the quantity in a given time and sums up?**

Ans. Integrating type instrument.

Q. **Name the different instruments according to the working?**

Ans.
1. Moving coil type instrument.
2. Moving iron type instrument.
3. Hot wire type instrument.
4. Dynamometer type instrument.
5. Induction type instruments.
6. Electrostatic type instruments.

Q. **What should be the properties of a good/ideal instrument?**

Ans. It should have dead beatness, applicability, should not consume more power, accurate, scale should be such that it may give quick vision to read, free from external atmospheric effects, easy to handle and pointer should be light enough.

Q. **Which is a deadbeat instrument?**

Ans. If the instrument rises quickly to its deflecting final position without oscillations the instrument is said to be deadbeat.

DIFFERENT FORCES EMPLOYED

Q. **What are the different forces employed in an instrument?**

Ans. These are three forces—the deflecting force, the controlling force and the damping force.

Q. **What do you mean by the deflecting force?**

Ans. The force which deflects the deflecting system is known as deflecting force or torque. Generally in all electrical instruments the force is nothing but the electrical quantity.

Q. **What is the controlling force?**

Ans. The force which controls the deflecting torque is known as controlling force.

Q. **Why controlling force is needed?**

Ans. To compare the electrical quantity with the calibrated mechanical force.

Q. **Will the moving system stop as soon as the controlling force is balanced?**

Ans. No, but it oscillates for some time.

Q. **What is the name of the force used to overcome the oscillations?**

Ans. Damping force.

Q. **Suppose the damping mechanism fails, what will happen?**

Ans. It will be difficult to note down the final reading quickly or say it will take more time to stabilise the reading because of oscillations.

Q. **How many methods are there for obtaining the controlling force?**

Ans. Gravity control and spring control.

Q. **Why the gravity control system is called so?**

Ans. Because the principle of gravitational pull is used in this controlling method.

Q. **What is the spring control method of controlling force?**

Ans. In this method, the spring is provided on the spindle and the force exerted depends upon the angle of twist or rewound.

Q. **Can you obtained a uniform, regular and accurate controlling force by one spring?**

Ans. No.

Q. **What are the drawbacks of one spring?**

Ans.
1. The change in temperature causes the change of length of the spring, which in other words affects the final reading.
2. The aging factor also changes stiffness constant of the spring.
3. One spring will not give the regular force and changes with the angle of torsion.

Q. **What is done to avoid the above errors?**

Ans. Double springs are used, wound in opposite direction, so that if one winds the other will unwind and a uniform torque results.

Q. **What should be the properties of the spring used for spring control?**

Ans. Uniform tension, low or negligible heating effect, non-magnetic and foolproof against mechanical fatigue.

Q. **What do you mean by the damping force?**

Ans. The force which is used to damp the oscillations is known as damping force.

Q. **What are the methods for obtaining the damping force?**

Ans. Air damping, fluid damping and eddy current damping.

Q. **Why damping device is essential in the instrument?**

Ans. To give quick and accurate final reading.

Q. **What is air damping?**

Ans. The damping obtained because of the air thrust.

Q. **How air damping is obtained?**

Ans. In the air damping, a chamber of aluminium has an aluminium vane, which moves with the deflecting system. The vane experiences a thrust in the opposite direction of the movement because of the oscillations and compression of air damps the oscillation.

Q. **What is eddy current damping?**

Ans. It is the most effective method. In this an aluminium or copper disc moves in the magnetic field so as to be cut by the line of force and eddy currents are set up in the disc. These currents set up a force opposing the cause (oscillations) responsible for the production of these currents and thus the oscillations are damped.

MOVING COIL TYPE INSTRUMENT

Q. **Why it is called a M.C. type instrument?**

Ans. Because a coil moves in the magnetic field so it is called as moving coil instrument.

Q. **What are the main parts of a moving coil instrument?**

Ans. Permanent magnet, coil, aluminium former, hair spring, pivots, spindle, pointer and calibrated scale.

Q. **How the coil is wound?**

Ans. Having more number of turns of fine wire.

Q. **How this current is fed to the coil?**

Ans. Through hair springs.

Q. **What is the material generally used for control spring?**

Ans. Phosphor bronze.

Q. **What type of damping is used in M.C. type instruments?**

Ans. Eddy current damping.

Q. **How eddy current damping is obtained?**

Ans. The aluminium or copper former is used. As this former rotates between the magnetic poles, the force is developed and provides a damping force to the oscillations.

Q. **How the deflecting torque and the current in the coil are related?**

Ans. The deflection is directly proportional to the current in the coil.

Q. **What is the nature of the scale, uniform or uneven?**

Ans. Uniform.

Q. **What are the uses of M.C. type instruments?**

Ans. It can be used as an ammeter and voltmeter.

Q. **Can you tell the connections if the m.c. type meter is used as an ammeter?**

Ans. Yes, it will be connected in series with the load.

Q. **Can a m.c. type meter be used as a voltmeter? If yes, how?**

Ans. Yes, can be used by connecting a resistance of suitable value in series with the instrument.

Q. **Can you use the m.c. type meter on A.C.?**

Ans. No.

Q. **If an ammeter is connected across the A.C. line what will happen?**

Ans. It will be damaged.

Q. **If you connect the voltmeter in series with the load will it burn out?**

Ans. No.

Q. **Which meter, the ammeter or voltmeter, has high resistance?**

Ans. Voltmeter.

Q. **Why a shunt is provided with an ammeter?**

Ans. To extend the range.

Q. **Name the device used for extension of voltmeter range?**

Ans. Potential diverter.

MOVING IRON TYPE INSTRUMENT

Q. **Why it is known as M.I. instrument?**

Ans. In these instruments, the moving part is a soft iron piece, so it is known as M.I. instrument.

Q. **What are the types of M.I. type instruments?**

Ans. Attraction type and repulsion type instruments.

Q. **What are the attraction type instruments?**

Ans. The M.I. instruments in which principle of magnetic attraction is applied are known as attraction type instruments.

Q. **Which magnetic property is used in repulsion type instrument?**

Ans. Repulsive, like poles repel each other.

Q. **What are the different parts of the M.I. instruments?**

Ans. In repulsion type instruments : Fixed coil, moving iron and fixed iron piece, spindle, pointer, calibrated scale, aluminium vane and chamber. In attraction type instruments : Fixed coil, aluminium vane and chamber, spindle, pointer, calibrated scale.

Q. **Where actually the repulsion takes place in case of repulsion type M.I. instrument?**

Ans. Between the soft iron pieces (placed in the same magnetic region) generally known as fixed and moving iron pieces.

Q. **How the scale of M.I. type instruments is calibrated?**

Ans. Uneven crowded in the beginning and end and spreaded in between.

Q. **Why the scale is uneven?**

Ans. Because the deflection is proportional to the square of the current flowing through the magnetic coil.

Q. **What type of damping is used in this type of instruments?**

Ans. Air damping.

Q. **Can this instrument be used for A.C. and D.C. both?**

Ans. Yes.

Q. **Up to what scale the reading of the instrument is reliable?**

Ans. The reading up to 1/10th of the full scale is not reliable.

Q. **Which instrument of M.I. type or M.C. type is robust?**

Ans. M.I. type.

Q. **Which value of current, r.m.s. or maximum, is measured by M.I. instruments?**

Ans. R.M.S. value.

Q. **What are the common errors in all instruments?**

Ans. Temperature error, frictional error and observation error.

Q. **What do you mean by the observation error?**

Ans. The error due to misreading the scale is known as observation error.

Q. **What is frictional error?**

Ans. The error due to the mechanical force acting at the pivots.

Q. **What is temperature error?**

Ans. The error is mainly due to heating of the working coil and other resistance connected in the instrument, apart from the room temperature, because of the operational current.

Q. **How the temperature error can be minimised?**

Ans. By using the coil and resistance of very less temperature coefficient materials.

HOT WIRE TYPE INSTRUMENT

Q. **Name the instrument in which heating effect of current is used?**

Ans. Hot wire type instrument.

Q. **Where actually heat is produced?**

Ans. In the wire which is connected between the fixed and adjustable screws.

Q. **What is the material used for this wire?**

Ans. Platinum iridium wire.

Q. **Is there any pulley used in the hot wire type instruments?**

Ans. Yes.

Q. **Name the wire used to transfer the sag of the platinum iridium wire?**

Ans. Phosphor bronze wire.

Q. **What for the spring is used in this instrument?**

Ans. To give motion to the pulley which results in the motion of the pointer.

Q. **What type of controlling is used?**

Ans. Gravity control.

Q. **What type of damping is used in this instrument?**

Ans. Eddy current damping.

Q. **How eddy current damping is used?**

Ans. Because the aluminium disc rotates between the magnetic poles of a permanent magnet, inducing eddy current and resulting the damping.

Q. **What is the nature of the scale?**

Ans. Uneven.

Q. **Why the scale is uneven?**

Ans. Here the deflection is because of the increase in length of platinum iridium wire, which is proportional to the square of the current, so deflection is proportional to the square of current, hence uneven scale.

Q. **What main precautions are to be observed in manufacturing of this instrument to avoid any error because of the increase in temperature or the atmospheric changes?**

Ans. The base of the instrument is made of the material of which coefficient of expansion is equal to that of hot wire, in order to minimise the error due to uneven expansion between wire and its fixture.

Q. **Why we use platinum iridium wire?**

Ans. Because the temperature coefficient is fairly constant and the working is safe up to 300°C.

Q. **Can you use this instrument on A.C. and D.C. both?**

Ans. Yes.

Q. **Why does it give correct reading on A.C. and D.C. both?**

Ans. Because the heat developed is proportional to the square of current and current on D.C. is equal to the r.m.s. value of current on A.C. side.

Q. **Is it advisable to use any safety device for the protection of overloading of this instrument?**

Ans. Yes, generally a fuse is provided.

Q. **Can you tell the very important utility of this instrument?**

Ans. It is not affected by any wave or frequency and stray magnetic effect so it can be used for any frequency.

Q. **Does the stray magnetic field affect the reading?**

Ans. Not at all.

Q. **How is the power consumption of the instrument?**

Ans. Power consumption is relatively high.

Q. **Once adjusted for zero position, will it be suitable for all the time to come?**

Ans. No, zero adjustment is required frequently.

Q. **Which thread is generally preferred between the spring and phosphor bronze wire?**

Ans. Silk thread.

Q. **In which position the instrument be kept for use?**

Ans. Vertically.

DYNAMOMETER TYPE INSTRUMENT

Q. **What is the working principle of this type of instrument?**

Ans. Same as that of motor; wherever a current-carrying conductor is placed in the magnetic field, a torque is developed on the conductor which tends to rotate it.

Q. **What type of coils are there in a dynamometer type instrument?**

Ans. Fixed coil and moving coil.

Q. **What do you know about fixed coil?**

Ans. It is generally wound into two halves, with thick conductors having less number of turns.

Q. **How the moving coil is wound?**

Ans. This coil has more number of turns of fine wire. Generally, it is air cored.

Q. **How moving coil is placed?**

Ans. It is fixed on the spindle so that it can move freely.

Q. **To avoid eddy current and hysteresis losses how the coils are wound?**

Ans. Air cored.

Q. **What type of controlling is used?**

Ans. Spring controlled.

Q. **Which type of damping is used?**

Ans. Air damping.

Q. **Is the scale even or uneven?**

Ans. Uneven.

Q. **Why the scale is not even?**

Ans. Because the deflection is again proportional to the square of current.

Q. **What are the uses of this instrument?**

Ans. It can be used as an ammeter, voltmeter and dynamometer type wattmeter, ampere hour meter, etc.

Q. **Can it be used for A.C. and D.C. both?**

Ans. Yes.

Q. **How the coils are connected while using as an ammeter?**

Ans. In series for small current but if the current is high then proper shunt is used.

Q. **How it is used as a voltmeter?**

Ans. The fixed and moving coils are connected in series but a suitable value external resistance is connected in series to drop down the excess voltage.

Q. **Can you use this instrument as a wattmeter? If so, how the connections are made?**

Ans. Yes, it can be used as a wattmeter. The fixed coil having less resistance is connected in series with the load and the moving coils having high resistance are connected across the line through a series resistance of suitable value.

Q. **What is the power?**

Ans. It is naturally the rate of doing work or the rate at which the energy is transferred. It is the product of voltage and current in D.C. circuit and in A.C. circuit the product of voltage, current and power factor.

Q. How many current coils are there in a single phase wattmeter?

Ans. Only one, divided into two halves.

Q. How many p.c. are there in a three phase wattmeter?

Ans. Three coils or sometimes two of 440 V.

Q. How the p.c. if three are connected?

Ans. In star.

Q. How many minimum c.c. and p.c. are possible for a three phase wattmeter even for unbalanced load?

Ans. Two pressure coils and two current coils.

Q. How many common errors are there in a wattmeter?

Ans. Error due to inductance of the pressure coil, due to capacitance of p.c. due to eddy currents, due to power loss in c.c. and p.c.

D.C. AMPERE-HOUR METER

Q. What do you mean by the D.C. ampere-hour meter?

Ans. It is D.C. energy meter.

Q. What are the different types of D.C. energy meters?

Ans. These are :

1. Ferranti mercury motor meter.
2. Elihu Thomson commutator motor meter.
3. Wound disc type meter.

Q. What are the different parts of the D.C. Ferranti mercury motor energy meter?

Ans. Fixed magnets, spindle, copper or aluminium disc, recording mechanism and mercury pool, etc.

Q. How many permanent magnets are there in this instrument?

Ans. Two in number.

Q. What are the functions of the two magnets?

Ans. One magnet is used for driving purpose and other magnet as a brake magnet for controlling purpose.

Q. How the damping force is produced?

Ans. By eddy current damping.

Q. If the current flowing through the meter is less a coil is provided to compensate for the driving, what is the name of the coil?

Ans. Compensating coil.

Q. What are the drawbacks of this meter?

Ans. 1. It cannot be used on A.C. and D.C. both.

2. If the part of the mercury is less or dirty, the instrument will not work satisfactorily.

Q. **How this meter is connected to the load?**

Ans. In series with the load.

Q. **How the current is lead in the instrument?**

Ans. Through one side by the copper brush to the copper disc and taken out from the pivot of the disc.

Q. **Can you tell how many purposes does the copper disc solve?**

Ans. Driving and controlling (half the disc for driving and other half for controlling).

Q. **Name the common errors in motor meter?**

Ans. The friction error and braking error.

Q. **Why it is called wound disc type meter?**

Ans. In these meters, there is a winding in between the discs or say the disc is wound.

Q. **How this instrument is connected in the circuit?**

Ans. In series with the load.

Q. **Name the constructional parts of the wound disc type instrument?**

Ans. Wound disc, spindle, jewelled bearing, recording mechanism, permanent magnet, commutator and brushes.

Q. **What are the drawbacks of the wound disc type instrument?**

Ans.
1. Cannot be used on A.C.
2. Requires special attention for heavy current operation.

Q. **How the current is passed through the wound disc?**

Ans. Through commutator and brushes.

Q. **If it is required to use the instrument on high current what arrangement should be done?**

Ans. Use the proper shunt.

Q. **Are the wound disc and Ferranti type meters the actual energy meters?**

Ans. These are simply the ampere-hour meters.

Q. **Name the meter in which voltage and current both are taken into account for energy metering?**

Ans. Elihu Thomson energy meter.

Q. **What are the constructional parts of the Elihu Thomson energy meter?**

Ans. Commutator, wound armature, current coils, aluminium disc, permanent magnet, spindle, recording mechanism, resistance and compensating winding.

Q. **How the magnetic poles are induced for driving purpose?**

Ans. By dividing the current coil into two halves.

Q. **What is the function of the compensating winding?**

Ans. If the load current is less and does not produce the sufficient driving torque, the compensating winding produces some flux to help.

Q. **How the compensating coil is connected?**

Ans. In series with the armature (pressure coil).

Q. **Why pressure coil circuit is made a revolving one?**

Ans. Because of the light weight. If the current coil is made revolving, the weight will be more and as a result power for turning will be more.

Q. **Has the operating torque any relationship with the current coil and armature circuit currents?**

Ans. Yes, it is proportional to the product of the currents flowing through the current coil and armature torque is proportional to the flux produced because of the current coil and pressure coil which is armature circuit in this case.

INDUCTION TYPE METER

Q. **Why these instruments are called induction type meters?**

Ans. Because these work on the principle of magnetic induction.

Q. **Are the instruments suitable for A.C. and D.C. both?**

Ans. Suitable for A.C. measurements only.

Q. **How much mechanical degree deflection is possible in these instruments?**

Ans. Upto 300°.

Q. **Name the main constructional parts of A.C. induction meter?**

Ans. The magnetic cores which are with the 'shaded' provision, spindle, controlling arrangement generally spring, aluminium disc, permanent magnet, pointer and calibrated scale.

Q. **Which damping method is adopted?**

Ans. Eddy current damping.

Q. **What do you understand by the exciting coil?**

Ans. It is the coil in which current flows and produces magnetic flux. In case of ammeter the number of turns of thick conductor are less and in case of voltmeter this coil is wound with thin conductor having more number of turns.

Q. **What is the function of aluminium disc?**

Ans. It serves two purposes :

1. It behaves as a short-circuited conductor and produces circulating current as a result the system is deflected.
2. Helps for eddy current damping.

Q. **What are the applications of induction type instruments?**

Ans. As ammeter, as a voltmeter and as a wattmeter etc.

A.C. ENERGY METER

Q. **Name the different energy meters generally used?**

Ans. Single phase energy meter, three phase energy meter.

Q. **What is the main constructional difference between the energy meter and wattmeter.**

Ans. The other construction is same except the control spring and pointer is replaced by the brake magnet, the aluminium disc and train of wheel for recording mechanism in energy meters.

Q. **How many main coils are there in an energy meter?**

Ans. C.C. and P.C. (current coil and pressure coil).

Q. **How C.C. is connected and wound?**

Ans. The current coil is connected in series with the load and wound with thick conductor having less number of turns.

Q. **What is p.c. and how it is wound?**

Ans. It is the pressure coil, it is connected across the line. It is wound with thin conductor having more number of turns.

Q. **What is the relationship between the current and voltage in pressure coil?**

Ans. The current lags behind the voltage by approximately 90°.

Q. **Name the provision which is made so that the angular displacement between the current and voltage is kept about 90° electrical?**

Ans. The shading ring is provided on the central limb of the cores.

Q. **How current and voltage relationship in the current coil is determined?**

Ans. It depends upon the nature of load. It will be same as that of the load current and voltage relationship.

Q. **What is the main operating mechanism or systems in induction type instruments?**

Ans. Driving system, moving system, braking system and registration system.

Q. **How many terminals are taken out from the energy meter?**

Ans. Four terminals, out of which generally central two terminals are short-circuited inside the meter.

Q. **Generally, where two terminals of c.c. are located?**

Ans. One and forth extreme two.

Q. **How pressure coil is connected?**

Ans. Across line and neutral. In the energy meter having four terminals, the coil is connected on the short-circuited terminal and the small screw near the current coil terminal.

Q. **Can you trace out the terminals of c.c. and p.c.?**

Ans. By means of testing lamp. The current coil terminals will give bright light and pressure coil only a small amount of sparking.

Q. **If the load on the meter is connected on the terminals of supply and supply on the load terminals, will the instruments give any reading?**

Ans. Yes, the reverse reading will be obtained.

Q. **What do you understand by the leg adjustment?**

Ans. An adjustment of flux and voltage to be in quadrature with the help of a shading bend, is known as leg adjustment.

Q. **What are the power factor, quadrature, inductive load adjustment?**

Ans. All are same. Only the different names are given. Same leg adjustment is done.

Q. **The shading bend sometime carries an additional resistance with a adjustable screw, why we use it?**

Ans. To improve the power factor error, so that the flux of p.c. is in quadrature to the voltage fed.

Q. **What are the errors generally experienced in energy meter?**

Ans. Phase and speed error, friction compensation, temperature and frequency error.

Q. **What do you understand by the term creeping?**

Ans. Sometimes a slow but continuous rotation is obtained even when the load is removed. This error is known as the creeping.

Q. **How overload compensation is done?**

Ans. Wherever the disc rotates in the field of permanent magnet, the eddy currents are induced causing self-braking. For the suitability of that magnetic force, the magnet is adjusted by providing the suitable magnetic shunt.

Q. **How many elements are there in a three phase four wire energy meter?**

Ans. Three.

Q. **Where, in case of balanced or unbalanced load, we use the 3-ϕ 4-wire energy meter?**

Ans. It can be used even with unbalanced load.

Q. **If a meter is running fast, how will you adjust the energy meter?**

Ans. In case of fast running the braking magnet is adjusted. If the magnet is brought towards the spindle the meter will run fast, if away from the spindle i.e., towards the periphery of the disc the meter will slow down.

ELECTROSTATIC TYPE INSTRUMENT

Q. **What is the principle of operation of the electrostatic type instrument?**

Ans. Attraction or repulsion between two charged bodies.

Q. **What is the actual formation of this instrument?**

Ans. Essentially a voltmeter.

Q. **What are the types of these instruments?**

Ans. Quadrant type, attraction type.

Q. **Where the quadrant type instruments are used?**

Ans. Voltage measurement up to 10 to 25 kV.

Q. **Where the attraction type instruments are used?**

Ans. Voltage measurement above 20 kV.

Q. **What are the main parts of the quadrant type voltmeter?**

Ans. Four fixed metal double quadrants, circular box, double ended vane, phosphor bronze wire or silvered quartz thread, damping arrangement and indicating arrangement.

Q. **How many discs are provided in an attraction disc type voltmeter?**

Ans. Two, mounted parallel to each other.

MEGGER

Q. **What is a megger?**

Ans. It is an instrument generally used for measuring the insulation resistance in megaohms.

Q. **How does it work?**

Ans. It is driven by means of a handle and generates the voltage which is used for measuring the insulation resistance.

Q. **What should be the speed of the generator?**

Ans. Generally 160 r.p.m. or as per the specifications mentioned.

Q. **For lighting load, which rating of the megger is used?**

Ans. 500 V.

Q. **What is the general ratio of voltage rating of the instrument and installation? How much should be the voltage of the instrument to be used for power installation testing?**

Ans. The rating is generally double than the voltage of the supply or appliance. For power installation, it should be of 1000 V megger.

Q. **What is the nature of the supply which is used in megger?**

Ans. D.C.

Q. **Can you allow the use of megger on the energised line or circuit?**

Ans. No.

Q. **What type of generator is used in this instrument?**

Ans. Permanent magnet type D.C. generator.

Q. **How many coils are there in the ohmmeter used in megger?**

Ans. Three, the current coil (deflecting coil) pressure coil and compensating coil.

Q. **How the pressure coil is connected?**

Ans. It is connected across the generated voltage through a suitable resistance to limit the current drawn.

Q. **How the current coil is connected?**

Ans. It is connected in series with the terminals, but even then there is a limiting resistance connected in series to limit the current in case of short circuit.

Q. **How many terminals are there in the instrument?**

Ans. E and L, two.

Q. **What for the ligaments are used?**

Ans. These are used to supply the current to the moving system.

Q. **Where does the pointer of the instrument rest?**

Ans. In between the range i.e. maximum and minimum readings.

Q. **What is the range of the scale?**

Ans. It is from zero to infinity.

Q. **If the terminals are short-circuited (E and L) what will be the reading of the instrument?**

Ans. Zero.

Q. **If the terminals E and L are open, what will be the reading of instruments?**

Ans. Infinity.

Q. **What are the advantages of compensating coil?**

Ans.
1. To cancel the effect of the external fields on the moving system.
2. To improve upon the division of scale.

Q. **What for the guards are provided?**

Ans. To shunt leakage current.

Q. **Will the aging factor of permanent magnet cause any effect over the performance of the instrument?**

Ans. No.

Q. **Why it cannot be affected?**

Ans. Because the instrument deflection is independent of the magnetic strength as the same magnets are used for the generating and motoring actions.

Q. **Does the toque developed in the megger depend upon the c.c. and p.c. or not?**

Ans. Yes, both coils.

Q. **How the graduations of the scale are made even or uneven?**

Ans. Uneven.

Q. **Does the generated voltage depend upon the handle speed?**

Ans. Up to certain limits. But if the speed is further fastened, there a clutch is fitted so that the speed and voltage remains constant if driven fast.

Q. **What care should be taken for the use of the megger?**

Ans.
1. Should in no case be used on energised line.
2. Before use check it first.
3. Handle should be rotated at 160 r.p.m. or the instructions supplied.

Q. **What is the general range of the instrument?**

Ans. 250 V to 2500 V.

Q. **Name the instrument used for finding out the earth resistance?**

Ans. Earth tester.

Q. **How many terminals are there in the earth tester?**

Ans. Four, P_1, P_2, C_1 and C_2.

Q. **How the connections of c.c. and p.c. are made?**

Ans. C_1, P_1 are short-circuited and are connected to earth electrode under test and P_2 and C_2 are connected to auxiliary electrode.

FREQUENCY METER

Q. **What is the frequency?**

Ans. It is the number of cycles per second.

Q. **What is the unit of frequency?**

Ans. c/s or Hz.

Q. **Name the instrument used for finding out the frequency?**

Ans. Frequency meter.

Q. **What are the types of the frequency meters?**

Ans.
1. Electrical resonance type.
2. Weston type.
3. Ratio meter type.

Q. **What is the mechanical resonance type meter?**

Ans. In this meter, the frequency say mechanical resonance is used, as every metal has its own natural (mechanical) frequency. So some thin strips called reeds are used. These are placed along with an electromagnet. The electromagnet produces the magnetic field of certain frequency then the reed having same natural frequency will start vibrating and the vibrations can be noticed by the flag mounted on one side of the reed.

Q. **Generally which size of the reeds are used?**

Ans. 4 mm width and 0.5 mm thickness.

Q. **Do you think that all reeds are identical?**

Ans. No, these are not identical.

Q. **How these differ?**

Ans. Slightly different in dimensions or carry different weight of the flag at their top.

Q. **How the vibrations are made to be efficiently noticeable?**

Ans. There is flag mounted over every reed which amplify vibrations.

Q. **When the mechanical reeds are used how their natural frequency is affected?**

Ans. The natural frequency can be affected or say vibrations are affected by the weight, length and size of the reed.

Q. **How the reeds are mounted?**

Ans. These are fixed on an armature, keeping the flags towards the face.

Q. **What is the unique advantage of this instrument?**

Ans. It gives reading free from error, due to change in temperature, wave form and the magnitude of the applied voltage provided the voltage is not too low to provide appreciable amplitude of the vibration.

Q. **What do you know about the energing coil wound for reed type instrument?**

Ans. It has more number of turns of fine wire and connected across the supply through a series resistance.

Q. **Can you take precise measurements of frequency by this reed type instrument?**

Ans. No, it is a disadvantage.

Q. **What is the electrical resonance frequency meter?**

Ans. The instruments in which the principle of electrical resonance is adopted are named as electrical resonance frequency meter.

Q. **How many coils are there in this instrument?**

Ans. Two—the fixed coil and moving coil.

Q. **What do you understand by the fixed coil in this instrument?**

Ans. It is the magnetising coil wound over the laminated iron cores and is so designed as to be connected across the line.

Q. **What is the shape of laminated iron cores?**

Ans. It is having a cross-section gradually varying section over its length being maximum near the end where the magnetising coil is mounted and minimum at the other end.

Q. **Name the constructional parts of this instrument?**

Ans. Magnetising coil, moving coil, pointer, pivots, laminated iron core and capacitor.

Q. **How the e.m.f. is induced in the moving coil?**

Ans. Because of the induction, as the flux of alternating nature is produced by the fixed coil.

Q. **How the scale is calibrated?**

Ans. It is having 50 c/s generally in centre and ranging from 45 c/s to 55 c/s in equal graduations.

Q. **What do you remark about the sensitivity of this instrument?**

Ans. In this instrument, the capacitance is of constant value and the inductance depends upon the position of the coil which is over the tapered laminations, the variable inductance causes a good sensitivity of the instrument.

WESTON FREQUENCY METER

Q. **What are the constructional parts of the instrument?**

Ans. Two pairs of the fixed coil set, long and thin soft iron needle, pointers, calibrated scale, three different inductances and two different resistances.

Q. **Why the fixed coil sets are made into two halves and are so arranged?**

Ans. The fixed coil sets are made into two halves and are so arranged as their axis are at right angle to the magnetic axis of the other coil.

Q. **What type of damping is used?**

Ans. There is no additional arrangement for damping provided. The magnetic inductance coil damps the oscillations.

Q. **What is the range of this instrument?**

Ans. It ranges from 47 c/s (Hz) to 53 c/s (Hz) generally for 50 c/s (Hz) frequency.

Q. **What is the advantage of this instrument?**

Ans. The scale can be extended to 300° mechanical i.e. wide range for each digit.

RATIOMETER TYPE FREQUENCY METER

Q. **Why it is called as ratiometer type frequency meter?**

Ans. Because it is working on the ratio principle.

Q. **What is the ratio-meter?**

Ans. The permanent ratio meter consists of two moving coils rigidly fixed together and mounted on the same spindle. Both coils move in the magnetic field produced by the permanent magnet.

Q. **On what factors does the deflection depend?**

Ans. The deflection depends upon the ratio of the two currents in the moving coils, hence name is also ratio meter.

Q. **What is the relationship between the current ratio and deflection of the ratio meter frequency meter?**

Ans. It is linear.

Q. **How the two coils are fed?**

Ans. By means of two separate bridge rectifiers.

Q. **What is the range of this instrument?**

Ans. It can be up to 5000 Hz.

POWER FACTOR METER

Q. **What is the power factor?**

Ans. It is the ratio of true power to apparent power.

Q. **If the angle of leads or lag is given how the power factor can be obtained?**

Ans. The power factor is the cosine of the angle of lead or lag.

Q. **What are the types of power factor meters?**

Ans. These are single phase and three phase.

Q. **What is the type of the meter (according to the working) the power factor meter?**

Ans. Electrodynamometer type.

Q. **How many coils are there in a single phase power factor meter?**

Ans. These are fixed coils and moving coils. The fixed coil is divided into two halves and the moving coils which are two, are placed at 90° to each other.

Q. **On what principle does the power factor meter work?**

Ans. On motor working principle i.e., whenever a current-carrying conductor is placed in the magnetic field a torque is developed which tends to move it.

Q. **How the supply is given to the moving coil?**

Ans. The pressure coils are the moving coils and are connected across the line, through the suitable values of resistance and capacitance in series, across the line.

Q. **What is the relationship between the current and voltage of pressure coil?**

Ans. The current in the coil, having capacitance in series, has current leading the voltage by approximately 90° and in the moving coil where resistance is connected in series, the current is in phase with the voltage.

Q. **How the current is related with the voltage in current coil?**

Ans. It depends upon the power factor of the load.

Q. **How the power factor of high voltage line is measured?**

Ans. By connecting the c.c. and p.c. of the power factor meter through current and potential transformers respectively.

Q. **How the torque is developed in case of power factor meter?**

Ans. When the meter is connected on the load current coils, causing the induction of the magnetic fields, which causes the turning movement by the interaction fields and finally the moving coils takes place the position.

Q. **What is the voltage rating of a dynamometer type power factor meter for balanced three phase load?**

Ans. 440 V 50 c/s.

Q. **How many minimum current coils does a power factor has for three phase balanced load?**

Ans. One which is divided into two halves.

Q. **For a three phase unbalanced load, how many coils are required?**

Ans. Three pressure coils and three current coils.

Q. **How the three pressure coils are connected for the three phase load?**

Ans. Generally in star.

Q. **How three current coils are connected?**

Ans. In series with the load.

Q. **How many electrical degree displacement does the current coils for three phase power factor has?**

Ans. 120° electrical.

SYNCHROSCOPE

Q. **What is the synchroscope?**

Ans. It is an instrument which is used for synchronizing two alternator with the busbar.

Q. **What are the conditions for synchronizing two alternator?**

Ans. The voltage, frequency, phase sequence must be same and voltage must be in phase.

Q. **What type of motor is used in the synchroscope?**

Ans. Split phase motor.

Q. **What are the different types of synchroscope?**

Ans. Dynamometer type and moving iron type.

Q. **If the frequency of incoming machine and busbar are same what will the pointer indicate on the scale?**

Ans. The pointer will stop in vertical position stationary or on the mark indicated.

Q. **If the synchroscope moves in anticlockwise position, what does it indicate?**

Ans. The incoming machine is slow.

Q. **What is the indication if the speed of the incoming machine is fast?**

Ans. Disc will rotate in clockwise direction.

Q. **How the synchroscope is used on high voltage line?**

Ans. With the instrument transformers.

Q. **How many windings are there?**

Ans. Two windings, one winding is fed by the incoming machine and the second winding is fed by the busbar or already running machine.

Q. **How many degrees of scale does this instrument has?**

Ans. 360°.

Q. **What is the use of this instrument?**

Ans. It is used for synchronising the alternators.

PHASE SEQUENCE INDICATOR

Q. **What do you understand by the phase sequence indicator?**

Ans. It is an instrument which indicates the sequence of the phases.

Q. **What is meant by the sequence of phases?**

Ans. In case of three phase system, the angular displacement is 120° elect, i.e. if one phase starts from 0°, the second phase start from 120° elect, and third phase from 240° elect, this arrangements of phases is known as phase sequence.

Q. **What is the working principle of the phase sequence indicator?**

Ans. Same as that of three phase induction motor.

Q. **How the windings are placed in this instrument?**

Ans. Three phase winding placed at 120° elect. phase displacement, generally connected in star and the remaining three ends are brought out for three phase supply.

Q. **How does it rotate?**

Ans. Because of induction of induced e.m.f. in the disc the eddy current causes the production of torque.

Q. **How the correct phase sequence is determined?**

Ans. If the disc rotates in the correct direction as marked then it is correct phase sequence and if not, i.e. in reverse direction then the phase sequence is not correct, change the connected leads.

POTENTIOMETER

Q. **What is the potentiometer?**

Ans. The potentiometer is essentially a device by which an unknown e.m.f. or potential difference either wholly or in part is compared against a standard source of e.m.f.

Q. **What is the standard reference source of voltage used for the potentiometer?**

Ans. Weston cadmium cell.

Q. **What is the order of accuracy obtained in the potentiometer?**

Ans. One part of 10^5.

Q. **How the e.m.f. is compared?**

Ans. In terms of length.

Q. **What is the e.m.f. of the standard cell?**

Ans. 1.0186 V.

Q. **What are the uses of the potentiometer?**

Ans. It is used for standardisation, measurement of unknown voltage and other accurate general purposes like measurement of resistance, power and current.

Q. **What are the errors and precautions for the potentiometer?**

Ans. The error can be in standard cell, standardisation, thermoelectric e.m.f., leakage and sensitivity of the potentiometer. It essentially requires the preventive maintenance regularly.

Q. **Can a potentiometer be used on A.C.?**

Ans. Yes, it is an A.C. potentiometer.

Q. **What are the applications of A.C. potentiometer?**

Ans. Calibration of voltmeter, calibration of ammeter, and the measurement of impedances, measurement of active and reactive power and measurement of wattage and energy.

Q. **What are the possible error of this instrument?**

Ans. Frequency error of the transformer, instrument, effects of harmonics in A.C. supply, phase angle error effects of inductive elements, effects of mutual inductance and effect of stray electric and magnetic fields.

OBJECTIVE TYPE MULTIPLE CHOICE QUESTIONS

1. While using an instrument for measurement or testing an electrical circuit, you are supposed to consider first ......
   (a) theoretical knowledge (b) practical knowledge
   (c) your surrounding (d) personnel safety
2. The permanent magnet moving coil type instruments are best suited for ......
   (a) A.C. measurement (b) D.C. measurement
   (c) A.C./D.C. measurement (d) frequency measurement.
3. The controlling method in m.c. type instrument is ......
   (a) spring control (b) gravity control
   (c) eddy current control
   (d) combination of spring and eddy current controlling.
4. The damping in m.c. type instruments is obtained by ......
   (a) spring damping (b) gravity damping
   (c) eddy current damping (d) friction damping.
5. The three main advantages of a m.c. type instruments are ......
   (a) well shielded, less expensive and suitable for A.C./D.C.
   (b) uniform scale, well shielded, suitable for A.C. only.
   (c) well shielded, more expensive and suitable for A.C. only.
   (d) uniform scale, high sensitivity and well shielded.
6. Moving iron type instrument are suitable for ......
   (a) D.C. measurements only
   (b) A.C. measurements only
   (c) A.C./D.C. measurements only.
   (d) resistance measurements only.
7. The damping in dynamometer type instruments is obtained by ......
   (a) air damping method
   (b) fluid damping method
   (c) eddy current damping method
   (d) spring control damping.
8. The electro-dynamometer type instruments can measure ......
   (a) A.C. only (b) D.C. only
   (c) A.C. voltage only.
   (d) A.C./D.C. voltage, current and power
9. Thermocouple type instruments are used for measuring ......
   (a) D.C. only (b) A.C. only

(c) A.C. and D.C. (d) static current and voltage.

10. All M.I. type instruments measure ......
(a) maximum value of A.C. (b) average value of A.C.
(c) r.m.s. value of A.C. (d) instantaneous value of A.C.

11. Heavy alternating current and voltage are measured by using ......
(a) shunt and multipliers (b) instrument transformers
(c) transformer and multipliers (d) current transformer and shunt.

12. The insulation resistance of the installation is measured by ......
(a) an ammeter (b) ohmmeter
(c) megger (d) kelvin bridge.

13. The postoffice box works on the principle of ......
(a) Kirchhoff's law (b) Wheatstone bridge
(c) Kelvin bridge (d) Maxwell's bridge

14. The instruments used for D.C. energy measurements are ......
(a) ammeter, voltmeter and galvanometer
(b) ammeter, ohmmeter and dynamometer type instrument
(c) Ferranti type, Elihu Thomson type and wound disc type instruments
(d) induction type, Elihu Thomson type and dynamometer type.

15. The wattmeter measures ......
(a) apparent power (b) true power
(c) reactive power (d) wattless power.

16. In case when the single phase energy meter is reading fast, the damping or holding magnet is moved ...... to slow down.
(a) towards the spindle.
(b) away from the spindle
(c) and removed from the meter
(d) and removed for a new replacement

17. According to the standard classification of the instruments these are ......
(a) ammeter, voltmeter and wattmeter
(b) moving coil moving iron and dynamometer type
(c) standard, substandard and commercial type
(d) frequency meter, power factor meter and energy meter.

18. The phase sequence indicator is used to indicate the ......
(a) frequency of the supply
(b) power factor of the supply
(c) energy consumed by the load
(d) phase sequence of A.C. polyphase supply.

19. The Wheatstone bridge gives correct reading if the galvanometer branch of the bridge is ......

(a) open circuit
(b) carrying some current
(c) carrying no current
(d) short-circuited.

20. A wattmeter has four terminals, while connecting to the load the fixed coil is put in ......
    (a) parallel with the load and moving coil in series with the load.
    (b) series with the load and moving coil in parallel with the supply.
    (c) series with the load both the coils.
    (d) parallel with the load both the coils.
21. Choose the correct answer for megger.
    (a) It is used to measure low/high/insulation resistance.
    (b) It has even/uneven scale.
    (c) It reads in MΩ/Ω.
    (d) It should be rotated at 160/320 r.p.m.
    (e) It will read zero/infinity, when connected across short-circuit.
    (f) It will read zero/infinity when connected across open circuit.
    (g) It has a hand-driven/electric-driven generator.
    (h) It has 500/1000 volt rating for domestic purposes.
    (i) It has 500/1000 volt rating for industrial use.
    (j) It is rated for low/high current rating.
22. Choose correct answer for M.I. type instruments.
    (a) It can be used for A.C./D.C. or A.C. or D.C.
    (b) It has even/uneven scale.
    (c) The 10/20% scale reading is not reliable.
    (d) It may be attraction/repulsion or both types.
    (e) It has two/three soft iron pieces in case of repulsion type.
    (f) It has air/eddy current damping.
    (g) The spring/gravity control is adopted in general.
    (h) It can be used as ammeter/voltmeter/ammeter and voltmeter.
    (i) It can be used as ammeter/wattmeter.
    (j) The scale can be extended upto 270°/90°.
    (k) The electromagnet or solenoid can be wound on the insulated/aluminium former.
    (l) The scale is crowded in the ends/middle.
23. Choose correct answer for electrodynamometer type instrument.
    (a) It can be used as ammeter/wattmeter/voltmeter/for all as given.
    (b) Generally eddy current/fluid damping is used.

(c) Current coil is made in two halves/three segments.

(d) The P.C. is made stationary/rotating.

(e) It has uniform/uneven scale.

24. An instrument indicates 25 A current in an induction motor line circuit. What values does it measure?

(a) Instantaneous (b) Average
(c) Effective (d) Peak.

25. Which meter is installed in the premises of a consumer for recovery of electrical energy?

(a) Ammeter (b) Power factor meter
(c) Energy meter (d) Voltmeter.

26. The single phase energy meter records the energy consumption in the unit of ......

(a) kiloampere hours (b) kilovolt hours
(c) kilowatt hours (d) None of the above.

27. In A.C. energy meter the braking force is developed by the ......

(a) load put on the meter (b) flux produced by A.C.
(c) flux produced by P.C.
(d) flux produced by permanent magnet

28. The eddy currents are produced in the ...... which is actually causing the rotation of disc.

(a) aluminium disc (b) current coil
(c) pressure coil (d) damping magnet

29. The holes in the aluminium disc help to reduce ......

(a) creeping (b) fast moving
(c) slow moving (d) the weight of the disc.

30. In A.C. single phase energy meter A.C. is connected in ......

(a) series with the load (b) parallel with the load
(c) neither in series nor in parallel.
(d) None of the above.

31. In A.C. single phase energy meter the pressure coil is connected in ......

(a) series with the load
(b) parallel with the supply voltage
(c) neither in series nor in parallel
(d) None of the above.

32. How many fluxes are in existence in the working A.C. single phase energy meter?

(a) only one of the current coils
(b) only two, one of current coil and second of the pressure coil.

(c) three, one of current coil, second of the pressure coil and third of the braking magnet.

(d) None of the above.

33. Which coil carries current proportional to the load current?

(a) Current coil
(b) Pressure coil
(c) Compensating coil
(d) None of the above.

34. Which coil carries current proportional to the voltage applied (supply voltage)?

(a) Current coil
(b) Pressure coil
(c) Compensating coil
(d) None of the above.

35. How the power factor error is corrected?

(a) By adjusting current coil.
(b) By adjusting pressure coil.
(c) By adjusting the power factor compensator.
(d) None of the above.

CHAPTER 20

Illumination

INTRODUCTION

Q. **Define illumination?**

Ans. The illumination may be defined as the light of proper colour, quantity and so directed as to allow quick and accurate vision without discomfort to the eyes.

Q. **What are the factors which affect the correct and accurate illumination?**

Ans. Nature of the work, architectural design, surroundings, kind of light and maintenance.

Q. **Which light is better for a job, facing to worker's eye or job?**

Ans. Facing to job, so that all light should fall on the object.

Q. **What are the characteristics of good illumination?**

Ans. Light should fall on job, correct size and type of lamp, proper location of the lamp and proper reflecting equipment.

Q. **What are the advantages of good illumination?**

Ans. Production will increase, less number of accidents and less wastage of the raw material.

TERMS AND DEFINITIONS

Q. **What is a standard candle?**

Ans. A standard candle has about 1/6 lb. weight of wax and burning at the rate of 7.77 gm/hr.

Q. **What is a standard candle power?**

Ans. It is the unit of illuminating power of a source. The one candle power is defined as the light coming out from a standard candle.

Q. **What do you call the ratio of illuminating power of a given source to that of the illuminating power of a standard candle?**

Ans. Candle power of that source.

Q. **What is light?**

Ans. It is defined as that radiant energy which produces the sensation of vision to the eyes.

Q. **What is luminous flux?**

Ans. It is the total quantity of light emitted by the source of light. Its unit is lumen.

Q. **What is lumen?**

Ans. It is the unit of luminous flux. It is the total quantity emitted per unit solid angle from a uniform source of one c.p.

Q. **What is the lux and foot candle power?**

Ans. The lux is otherwise called as metre candle. It is the illumination produced on the inner surface of a sphere of one metre radius by a point source of one c.p. placed at its centre.

The foot candle is also defined as the illumination produced on the inner surface of a sphere of one foot radius by a point source of one c.p. placed at its centre.

Q. **What will be the total flux emitted by a source of 40 c.p.?**

Ans. It is $4\pi I = 4 \times 3.14 \times 40 = 502.6$ lumen.

Q. **What do you know about MSCP and MHSCP?**

Ans. The m.s.c.p. is the mean spherical candle power which is defined as average candle power of a source in all directions,

i.e. $$\frac{\text{Total flux in lumens}}{4\pi}$$

and m.h.s.c.p. is the mean hemispherical candle power and is defined as the average candle power over a hemisphere.

i.e. $$\frac{\text{Total flux emitted in lumens}}{2\pi}$$

Q. **What do you call the ratio of m.s.c.p. and m.h.s.c.p.?**

Ans. Reduction factor.

Q. **What is light efficiency?**

Ans. The ratio of lumen output to wattage taken is termed as light efficiency.

Q. **What do you say about the life of the lamp?**

Ans. It is the total number of hours operated giving light before burnt out.

Q. **What is brightness and its unit?**

Ans. It is defined as the flux emitted per unit area of the source in direction perpendicular to the surface. The unit of brightness is lumens per square metre.

Q. **What are the laws of illumination?**

Ans. These are :

1. *Inverse square law.* The illumination on a surface varies inversely as the square of the distance from the source of light.
2. *Lambert's cosine law.* The illumination varies directly as the cosine of the angle between the normal to the surface and direction of incident light.

$$E = \frac{1}{d^2} \cdot \cos \theta$$

Q. **Do you know about instrument which can directly measure the light?**

Ans. Yes, it is autophotometer.

Q. **What are the different kinds of light?**

Ans. Direct light, indirect light and semi-direct light.

Q. **How much percentage of light is reflected upwards in case of indirect lighting?**

Ans. 90%.

Q. **Tell some suitable values of illumination for corridor, living room, offices, drawing office and main road?**

Ans. 20-30 lm/m^2, 50-60 lm/m^2, 100-120 lm/m^2, 250-400 lm/m^2, 5-10 lm/m^2, respectively.

Q. **What is a flasher and what are its types?**

Ans. Flasher is a drum switch and these are used to produce the lighting effect. These are rotating drum type and automatic time contactor type.

Q. **What is the depreciation factor?**

Ans. It is the ratio of illumination under normal working conditions to the illumination under ideal conditions when lamp is new and cleaned also.

Q. **Is there any relationship between the visible radiant energy and temperature of radiant body?**

Ans. Yes, the visible radiation is directly proportional to the temperature of the body, less temperature less radiation and more temperature more radiation.

ELECTRIC LAMPS

Q. **What is the electric lamp?**

Ans. It is the source which changes the electrical energy into heat and then to the visible radiation i.e., light.

Q. **What are the types of lamps?**

Ans. Incandescent lamp and gas discharged lamp.

Q. **Define incandescent lamp?**

Ans. The lamps in which light is coming out directly by heating of the element are known as incandescent lamp.

Q. **What are the gas discharged lamps?**

Ans. The lamp in which light is radiated by the discharge through the luminous material.

Q. **Give the example of incandescent and gas discharge lamp?**

Ans. Incandescent lamps—Ordinary filament lamps.

Gas discharge lamps—Fluorescent tube, HPMV lamp, sodium vapour lamp and neon sign etc.

Q. **The incandescent lamps are of two types—one is the gas-filled incandescent lamp, name the other type?**

Ans. Vacuum type lamp.

Q. **What is the main function of gas in the gas-filled lamp?**

Ans. The gas is the inert gas and does not allow the oxidation of the filament.

Q. **What are the type of incandescent lamps?**

Ans. These are carbon filament lamp, metal filament and carbon arc lamp.

Q. **What is the type of carbon filament lamp and where these are used?**

Ans. It is a type of incandescent lamp and these are used for voltage drop wherever required, for example in battery charging.

Q. **What type of temperature coefficient does the carbon has?**

Ans. Negative temperature coefficient.

Q. **What is the working temperature of the carbon filament and why these are not in common use?**

Ans. The working temperature is 1800°C. At higher temperature carbon evaporates, as a result the container becomes black and life is reduced because of negative temperature coefficient.

Q. **What is the approximate life of the carbon filament lamp?**

Ans. 800-880 hrs.

Q. **Which material is used for making filament of the incandescent lamp?**

Ans. Tungsten.

Q. **What is the melting temperature and working temperature of the tungsten?**

Ans. 3500°C (approx.) and 1850°C, respectively.

Q. **If you increase the temperature of the lamp filament will it have any effect on the efficiency?**

Ans. The efficiency of the lamp could be increased.

Q. **What will be the limit of temperature?**

Ans. The limit of temperature will be the melting point, but the safe current limit is well below this, because the high temperature results short life.

Q. **Can you tell the radiant energy will be in which form of regions?**

Ans. High temperature of filament results a small portion of emitted energy as visible light, very little given out in ultraviolet region but most of the energy is given out in infra-red region.

Q. **Which type of spectrum is available by an incandescent lamp?**

Ans. It is a continuous spectrum, having light of every colour having no gap in radiant energy.

Q. **What type of filaments are used?**

Ans. Straight and coil coiled.

Q. **Is there any advantage of using coil coiled filament?**

Ans. Yes, in a small portion the length of the filament is concentrated which helps of producing more temperature effectively.

Q. **If a filament is directly connected to the supply voltage in open atmosphere, what will happen?**

Ans. The oxidation of filament will take place. White fumes will be seen and filament will burn out.

Q. **What is done to avoid the oxidation?**

Ans. The lamps are made either vacuum type or gas-filled type.

Q. **What are the parts of the incandescent lamp?**

Ans. Cap, glass envelope, filament, filament supports, stud and stump rod and exhaust tube.

Q. **Can you tell the different caps used for incandescent lamp?**

Ans. Bayonet cap, screw cap (the screw caps are of Edison type and Goliath type).

Q. **What is the unique characteristic of the bayonet cap on incandescent lamp?**

Ans. The pins of the cap and connection strips (pin) are at 90° to each other.

Q. **Sometimes it is observed that in spite of being everything alright if the lamp is inserted in the holder it does not light, why?**

Ans. Apart from all electrical faults, the important point is that the pins to lead the current in and out and the cap supporting grooves are not at right angle to each other, check and correct them.

Q. **Can you tell the diameter of the cap?**

Ans. 22 mm.

Q. **What lengths of the cap are in use?**

Ans. It varies from 21 mm to 31 mm.

Q. **Which type of cap is used for 40 W, 60 W and 500 W lamps?**

Ans. 40 W, 60 W, Bayonet cap and 300 W and 500 W screw cap.

Q. **Which shapes of glass are in the market for incandescent lamp?**

Ans. Pearly Pygmy, candle, twisted candle, flame tubler.

Q. **What is the use of exhaust tube?**

Ans. It is the tube used for exhausting purpose and has filling purpose after that the tube is sealed.

Q. **How much voltage is used for D.C. arc lamp?**

Ans. 42-50 V.

Q. **Which electrode is thicker in D.C. arc lamp?**

Ans. Positive.

Q. **Why positive electrode is made double in cross-section?**

Ans. Because the rate of consumption of positive electrode is double than the negative electrode.

Q. **How much is the length of arc?**

Ans. 1/8″-1/4″.

Q. **Can we use arc lamp on A.C. supply?**

Ans. Yes.

Q. **What is the ratio of light emitted from the arc lamp?**

Ans. D.C. arc lamp—85% light from positive and 10% from negative electrode.
A.C. arc lamp—50% from phase and 50% from neutral electrode.

Q. **What is the working voltage across A.C. arc lamp?**

Ans. 50-60 V approx.

Q. **What is the fundamental principle of gas discharged lamp?**

Ans. Ionization of gases.

Q. **Why discharge lamps are so popular?**

Ans. Because of light efficiency, uniform intensity of light and more life etc.

Q. **What are the constructional part of fluorescent tube?**

Ans. Glass tube, filaments, pins, powder for coating and gas to be filled.

Q. **Why fluorescent tube is so popular?**

Ans. Because of its cool and uniform light, high efficiency, suitable for A.C. or D.C. both, can be used in any direction and position.

Q. **What are the accessories required for tube connections.**

Ans. Tube, starter, choke, holder, wires and strip.

Q. **What do you know about choke?**

Ans. It is ballast. It has laminated cores, which are wound with thin conductors having more number of turns. It gives high voltage at the time of starting the tube and drops down the voltage in running condition.

Q. **What are the functions of a choke?**

Ans.
1. To give a surge voltage at the time of starting of the tube.
2. To drop the excess voltage.
3. To limit the current.

Q. **While purchasing a choke for a certain tube, what care should be taken?**

Ans. The wattage of choke and tube must be same.

Q. **What is the function of a starter?**

Ans. To start the tube. It is used to make and break the continuity so that the choke may induce high voltage for starting the tube.

Q. **What are the types of tube starters?**

Ans. Glow type and thermal type.

Q. **What are the parts of glow type starter?**

Ans. Aluminium or plastic container, two bimetallic strips glass tube, condenser and gas (He + Neon).

Q. **At the time of staring how much voltage acts across the starter?**

Ans. 220 V. approx. or the supply voltage.

Q. **In starting condition and running conditions the same connections remain intact, why in running conditions the starter does not make and break the contact?**

Ans. In running condition of the tube, only about 110 V or half the voltage works across the starter. That voltage will not be sufficient to produce spark between the strips, so in that condition it remains almost isolated.

Q. **What is the basic difference between the glow type and thermal type tube starters?**

Ans. In thermal type starter, a heater element is connected in the starter. Secondly in this starter both bimetallic strips are having normally close contacts. As the current starts flowing the bimetallic strips will bend and make an open circuit, whereas in case of glow type starter the contacts are normally open.

Q. **In fluorescent tube why light efficiency is high?**

Ans. In the case of these lamps the invisible radiation is also converted into visible radiation because of the phosphor coating inside the tube thus efficiency is increased.

Q. **What is meant by HPMV lamp?**

Ans. It is high pressure mercury vapour lamp.

Q. **How many times does a HPMV lamp give more light than that of an incandescent lamp?**

Ans. It gives approximately 2.5 to 3 times more light than an incandescent lamp of the same wattage.

Q. **How many electrodes are there?**

Ans. Two main electrodes and one auxiliary electrode.

Q. **What is an auxiliary electrode?**

Ans. It is an electrode having high resistance mounted near the electrode of opposite polarity.

Q. **What is the function of the auxiliary electrode?**

Ans. Just to ionize the gas at the time of starting of lamp.

Q. **How much is the pressure in the inner tubes of the HPMV lamp?**

Ans. It is about 1½ to 2 atmospheric pressure.

Q. **Why the space between the two tubes is evacuated?**

Ans. To reduce the heat loss.

Q. **Does the auxiliary electrode serve any purpose?**

Ans. It helps in starting and after that there is no function of the electrode.

Q. **Which gas is generally filled in inner tube?**

Ans. Argon with small amount of mercury.

Q. **What types of mercury vapour lamps are there?**

Ans. MA, MAT, MB types.

Q. **What is MA, MAT and MB type lamps?**

Ans. MA—Mercury vapour lamp with auxiliary electrode.

MAT—Mercury vapour lamp with auxiliary electrode and tungsten filament.

MB—Mercury vapour lamp with bayonet cap.

Q. **Is there any quartz in the 'MB' type lamp, if yes, why?**

Ans. Yes, the lamp generally contains the quartz which helps in withstanding the high temperature, hence it can be used in any direction.

Q. **Which colour light does the sodium vapour lamp give?**

Ans. Yellow.

Q. **Which out of two lamps has more length?**
(i) Sodium vapour lamp (ii) H.P. M.V. lamp.

Ans. Sodium vapour lamp.

Q. **Which gas is filled in sodium vapour lamp?**

Ans. Neon with some sodium drops.

Q. **To establish the discharge, how much voltage is required in sodium vapour lamp of 100 watts?**

Ans. 450 volts (approx.).

Q. **How you are getting this voltage for starting the sodium vapour lamp?**

Ans. By using high leakage reactance transformer.

Q. **What is the working temperature of this lamp?**

Ans. About 300°C.

Q. **Which lamp takes more time to give full light?**
(i) H.P.M.V. lamp (ii) Sodium vapour lamp.

Ans. Sodium vapour lamp.

Q. **Can you tell the approximate working temperature of the HPMV lamp?**

Ans. 600°C.

Q. **Where these lamps are used?**

Ans. For decoration purpose, street lighting, etc.

Q. **Can you tell the different wattages of the HPMV lamp, sodium vapour lamp and fluorescent tube?**

Ans. HPMV lamp—60 W, 125 W, 250 W, 400 W.
Sodium vapour lamp—40 W, 90 W, 100 W, etc.
Fluorescent tube—20 W, 30 W, 40 W, 80 W, 125 W.

Q. **What is the length and diameter of 80 W, 125 W, 20 W and 40 W tube?**

Ans. 80 W—5′ and 1½″, 125 W—8′ and 1½″, 20 W—2′ and 1″ and 1½″, 40 W—4′ and 1½″.

Q. **Name the lamp in which ultraviolet radiation principle is used?**

Ans. Ultraviolet lamps.

Q. **Where the neon signs are used?**

Ans. For advertisement and indicating purpose.

Q. **Name the bulbs which come under neon signs?**

Ans. Neon bulbs, line tester, osglim lamp, neon signs tubes, etc.

Q. **How different colours are produced?**

Ans. With the combination of gases.

1. For orange or red colour—Neon gas
2. Blue colour—Mercury in neon gas
3. Green colour—Yellow glass and mixture of neon gas with mercury
4. For yellow colour—Yellow glass with helium.

Q. **Is there any filament used in these bulbs?**

Ans. No, only electrodes are used.

Q. **In neon lamps generally which value of resistance is connected in series?**

Ans. 2000 Ω.

OBJECTIVE TYPE MULTIPLE CHOICE QUESTIONS

1. The phenomenon of producing the sensation of vision to the eyes because of the radiant energy is known as ......
   (a) light (b) sound (c) heat (d) electricity.
2. Lumen is the unit of ......
   (a) luminous flux (b) electric flux
   (c) magnetic flux (d) None of the above a, b and c.
3. The spherical reduction factor is the ratio of ......
   (a) r.m.s. value to average value of A.C.
   (b) m.s.c.p. and m.h.c.p.
   (c) m.s.c.p. to flux produced
   (d) m.h.s.c.p. to average value.
4. "The illumination on a surface varies inversely as the square of distance from the source of light" is the ......
   (a) cosine law (b) Coulomb's law
   (c) inverse square law (d) Ohm's law.
5. The lamps in which light is directly coming out by the heating of the filament are known as ......
   (a) discharge lamp (b) neon lamps
   (c) incandescent lamp (d) fluorescent lamps
6. The mercury vapour lamps are ......
   (a) M.A., M.A.T. and M.B. type
   (b) M.A., A.B. and M.A.B.T. type
   (c) M.A., M.A.T. and A.B.T. type
   (d) Bayonet screw and M.B. type
7. The filament of incandescent lamp is commonly made of ......
   (a) carbon (b) tungsten (c) nickel (d) nichrome.
8. Name the apparatus used for starting a fluorescent tube ......
   (a) choke (b) tube holder
   (c) starter (d) None of the above.
9. The function of the choke in the fluorescent tube circuit is ......

(a) to limit the current to safe valve
(b) to limit the current to safe value and give surge voltage for the start of discharge
(c) to increase more light output
(d) to supply a surge voltage for starting the discharge.

10. The ideal material for the filament of incandescent lamp is one which has ......
(a) high melting and operating temperature
(b) low vapour pressure, ductile and mechanically strong
(c) high specific resistance and low temperature coefficient
(d) high melting and operating temperature, low vapour pressure, ductile and mechanically strong, high specific resistance and low temperature coefficient.

11. The main object of vacuum type incandescent lamp is ......
(a) to prevent oxidation
(b) to minimise loss of heat by convection
(c) to check vaporisation of filament material
(d) to prevent oxidation, minimising loss of heat by convection and to check the vaporisation of the filament material.

12. What is the approximate value of auxiliary electrode resistance placed in series for starting in H.P.M.V. lamp ......
(a) 10 kΩ (b) 5 kΩ (c) 50 kΩ (d) 1 kΩ.

13. In H.P.M.V. lamp choke is used ......
(a) to limit the current to safe value
(b) to give surge voltage for establishing discharge
(c) to drop down the voltage in starting
(d) not to interrupt the lamp circuit.

14. In sodium vapour lamp the high leakage reactance transformer is used which gives about ...... volts in starting.
(a) 450 V (b) 110 V (c) 1000 V (d) 2000 V.

15. In discharge lamps, the fluctuation in light output which causes multiple images appearance of moving objects with jerks and reversed direction is called ......
(a) heating effect (b) stroboscopic effect
(c) lighting effect (d) X-rays effect.

16. A simple apparatus called ...... is used for the measurement of candle power.
(a) ammeter (b) voltmeter
(c) photometer (d) fluxometer.

17. The beam factor which is the coefficient of utilisation generally lies between ......

    (a) 0.5 to 1 (b) 0.3 to 0.5
    (c) 1 to 1.5 (d) 10 to 15.

18. Which lamp gives nearly a monochromic light.

    (a) Tube light (b) Sodium vapour lamp
    (c) Incandescent lamp (d) Kerosene lamp.

19. The zero watt lamp consumes about ......

    (a) zero watt (b) 5 to 7 W
    (c) 30 to 40 W (d) 0.5 to 0.8 W.

20. The general height of street lights is ......

    (a) 2 m (b) 3 to 4 m (c) 20 m (d) 10 to 15 m.

CHAPTER 21

Converting Machines

CONVERSION OF A.C. INTO D.C.

Q. **What is the necessity of conversion A.C. to D.C.?**

Ans. Nowadays about 90% of the electricity generated is of alternating nature. There are certain places where D.C. is essentially required like electroplating, battery charging, electrolysis, etc. So for that purpose the conversion is essential.

Q. **Which supply is effective for controlling the relay and speed control?**

Ans. D.C. supply

Q. **What are converters and inverters?**

Ans. The machine which converts A.C. into D.C. in known as converter and the machine which converts D.C. into A.C. is known as the inverter.

Q. **What are the methods of converting A.C. into D.C.?**

Ans. MG set, motor converter, rotary converter, mercury arc rectifier and metal rectifier.

Q. **What is MG set?**

Ans. The set which has D.C. generator and A.C. motor coupled together and converts A.C. into D.C. is known as MG set.

Q. **What actually is changed in MG set?**

Ans. A.C. energy is changed into mechanical energy and that mechanical energy obtained from the A.C. motor is converted into D.C. energy by D.C. generator.

Q. **Can you control the D.C. output?**

Ans. Yes, by controlling the field excitation of the D.C. generator.

Q. **The MG set is a self-starting one or not?**

Ans. Yes, it is a self-starting one.

Q. **What are the advantages of MG set?**

Ans.
1. The D.C. output is practically constant.
2. D.C. can be controlled easily.
3. Set is self-starting.
4. Can also be used for power factor correction.

Q. **What are the disadvantages of M.G. set?**

Ans.
1. It is heavy.
2. Requires more space.
3. Two machines are there, so maintenance is more.
4. Comparatively less efficiency.
5. Cost is more.

Q. **Which A.C. motor is generally used in MG set?**

Ans. A.C. squirrel cage induction motor.

Q. **Can it be coupled with the belt?**

Ans. Yes, but the efficiency will be further less.

Q. **How generally the squirrel cage induction motor of M.G. set is started?**

Ans. By applying low voltage at the time of starting by means of the autotransformer or star-delta starter and then it is put across the line.

MOTOR CONVERTER

Q. **What is a motor converter?**

Ans. It is essentially an ordinary slipring motor coupled both electrically and mechanically to D.C. generator.

Q. **How many phase windings it has?**

Ans. Generally 12 phase winding.

Q. **How the rotor winding is connected?**

Ans. In star under normal operating conditions.

Q. **How the input to the motor converter is fed?**

Ans. Partly by mechanically (through shaft) and partly electrically.

Q. **What are the drawbacks of the motor converter?**

Ans.
1. Low speed operation.
2. It is more expensive, hence not in common use.

ROTARY CONVERTER

Q. **What is the difference between rotary converter and synchronous converter?**

Ans. Both are same.

Q. **What are the constructional parts of the rotary converter?**

Ans. It is a machine, having armature and field winding. On one side of the armature there are sliprings and on the other side the commutator is installed. The armature winding is connected to both systematically.

Q. **D.C. is obtained from which side?**

Ans. From commutator side.

Q. **Is it a self-starting machine?**

Ans. Yes, if started from A.C. side as a polyphase induction motor and that is because of damping winding.

Q. **Let us say that the brushes are lifted, name the machine the rotary converter behaves if it is started from A.C. mains?**

Ans. Synchronous motor.

Q. **If the sliprings are disconnected, name the motor the converter behaves.**

Ans. D.C. shunt motor.

Q. **Can you take A.C. and D.C. simultaneously from a single machine?**

Ans. Yes.

Q. **How will you obtain A.C. and D.C. both?**

Ans. If the armature of converter is driven by the separated source, then A.C. side will give A.C. voltage and from commutator we can have D.C. voltage.

Q. **How A.C. is fed to the converter?**

Ans. Through sliprings.

Q. **If the rotary converter is to be started from D.C. side, where it is fed?**

Ans. Through brushes to the commutator.

Q. **Can you give any other name than the rotary converter, keeping in view the operation i.e., converting from A.C. to D.C. to A.C.?**

Ans. While it is converting from A.C. to D.C. it is called converter and when D.C. to A.C. it is named as inverter.

Q. **How many sliprings are there in a single phase converter?**

Ans. Two.

Q. **Where lap and wave windings are preferred?**

Ans. Wave winding is suitable for small machines and that is why it is not commonly used. Lap winding is commonly used in most of the large machines.

Q. **What are the equipotential points?**

Ans. The points in the winding where the e.m.f. induced has same potential.

Q. **What will happen if the equipotential points are short circuited?**

Ans. Nothing will happen, because the points are having zero potential difference.

Q. **In a lap winding how many tappings are taken to the sliprings?**

Ans. The number of tappings are equal to the product of number of sliprings and the number of pair of poles.

Q. **How many tappings are taken to sliprings of a three phase six pole synchronous converter?**

Ans. The tappings are equal to the number of sliprings multiplied by the number of pair of poles.

It is a three phase machine so slipring are three.

$$\therefore \quad \text{Tappings} = 3 \times \frac{6}{2} = 9$$

Q. **How many sliprings are there in a single phases, three phase, six phase, twelve phase converter?**

Ans. 2, 3, 6 and 12, respectively.

Q. **What will be the angular displacement between the adjacent tapping in a three phase, twelve phase synchronous converter?**

Ans. 120° elect., 30° elect, respectively.

Q. **Can you tell the formula for finding out the angular displacement between the adjacent tappings?**

Ans. $2\pi/m$, where m is the number of sliprings.

Q. **What is the voltage ratio between A.C. (between the sliprings) and D.C. output voltage?**

Ans. $E_{ac} = \dfrac{E_{dc} \sin \pi/m}{\sqrt{2}}$

Q. **What is the voltage relationship for three phase rotary converter?**

Ans. $E_{ac} = \dfrac{E_{dc} \sin 60}{\sqrt{2}} = 0.612 E_{dc}$

(E_{ac} = voltage at sliprings; E_{dc} = D.C. voltage)

Q. **What is the current relationship between the A.C. and D.C.?**

Ans. $\dfrac{I_{ac}}{I_{dc}} = \dfrac{2\sqrt{2}}{\eta\, m \cos \phi}$

where I_{ac} = line current on A.C. side; I_{dc} current on D.C. side; η = efficiency, m = number of phases, $\cos \phi$ = power factor.

Q. **What is the current relationship of single phase rotary converter if $\eta = 100\%$ and power factor is unity?**

Ans. $I_p = 1.41 I_{dc}$

Q. **What is the current relationship between A.C. and D.C. for three phase converter having 100% efficiency and unity power factor?**

Ans. $I_p = 0.545 I_{dc}$

Q. **What are the relationships of current and voltage of six phase supply?**

Ans. $I_{ac} = 0.472 I_{dc}$

$V_{ac} = 0.354 V_{dc}$

Q. **What are the methods of connecting the six sliprings of the converter?**

Ans. These are in double star, double delta or diametrical connections.

Q. **What are the methods of starting a rotary converter?**

Ans. Direct starting from A.C. side by means of :

1. By taps on the main transformer;
2. By separate starting motor;
3. By self-synchronizing method.

Q. **What precaution is must while starting from D.C. side?**

Ans. A.C. side must be open and sliprings be isolated.

Q. **Is it necessary to disconnect D.C. side while starting from A.C. side?**

Ans. Yes.

Q. **The main poles of synchronous converter are slotted and provided with the special winding. Name the winding and the use of that winding.**

Ans. Damper winding. It is used for starting the converter and hunting the oscillations.

Q. **Which motor is generally used for starting by separate means?**

Ans. Squirrel cage induction motor direct coupled to the shaft and having less number of poles than the converter.

Q. **The ratio of A.C. and D.C. voltage is fixed in any converter, so can you control the D.C. voltage?**

Ans. Yes, by changing the applied A.C. voltage.

Q. **What are the methods for changing the D.C. voltage in the rotary converter?**

Ans.

1. Series reactance control.
2. By changing tap changers on the supply of the input.
3. Induction regulator.
4. Synchronous booster control.

Q. **Can the rotary converter be subjected to parallel operation, if the load demand is more?**

Ans. Yes, can be put in parallel operation.

Q. **Which side of the converter is operated for parallel operation?**

Ans. D.C. side as shunt or compound generator.

Q. **Which characteristic is essential to observe for successful operation?**

Ans. Drooping external characteristic.

Q. **What is the limitation of the voltage?**

Ans. 1200-1500 volts.

Q. **What are the advantages of the rotary converter?**

Ans. Can be used to convert A.C. into D.C. or D.C. into A.C., good efficiency, reliable operation, less space required, less heating effect and maintenance cost is also less.

RECTIFIERS

Q. **What is rectifier?**

Ans. It is a static device, which is used to convert A.C. into D.C.

Q. **What are the different types of rectifiers?**

Ans. Hot cathode rectifier, mercury arc rectifier and metal rectifier.

Q. **What is meant by cathode?**

Ans. Cathode is that which emits electrons or from where the electrons are emitted.

Q. **What is anode?**

Ans. The anode is a plate which attracts the emitted electrons.

Q. **What do you understand from the half-wave rectification?**

Ans. If the rectifier converts only half the cycle of A.C. into D.C. the process is known as half-wave rectification.

Q. **What is full-wave rectification?**

Ans. If the full wave of A.C. is converted into D.C. the process is called full-wave rectification.

Q. **What are the types of mechanical rectifier?**

Ans. These are two—the commutating type rectifier and vibrating rectifier.

Q. **What is the type of Tunger bulb rectifier, whether hot cathode or cold cathode?**

Ans. Hot cathode rectifier.

Q. **For low capacity Tunger bulb rectifier the cathode is made of which material?**

Ans. Tungsten.

Q. **Why the cathode is oxide coated?**

Ans. To emit the electrons efficiently.

Q. **Generally which material is used for anode?**

Ans. Barium.

Q. **In a half-wave rectifier how many anodes and cathodes are used?**

Ans. One anode and one cathode.

Q. **How many anodes and cathodes are used for full-wave rectification?**

Ans. One cathode and two anodes for single phase rectifier.

Q. **How the cathode is energised?**

Ans. The cathode is energised by the separate winding.

Q. **Where this separate winding is placed?**

Ans. On the main transformer and designed for low voltage.

Q. **What is the polarity of cathode?**

Ans. It emits the electrons so for the external circuit it behaves like positive terminal.

Q. **The anode can be made of cylindrical cast iron, is it? If yes, upto which capacity?**

Ans. Yes, for big capacity of rectifiers.

Q. **Why mercury vapours are added?**

Ans. To increase the current capacity.

Q. **Name the hot cathode rectifier consisting of glass envelope containing argon at reduced pressure and enclosing two electrodes?**

Ans. Hot cathode gaseous rectifier.

Q. **What is the utility of gaseous rectifier?**

Ans. In the gaseous rectifier, a large number of ions can be produced with the same potential difference between cathode and anode than the ordinary hot cathode rectifier.

Q. **Can you give the voltage drop in the Tunger bulb rectifier?**

Ans. It varies from 6 to 40 volts depending upon the size and capacity of the bulb.

Q. **What is the approximate efficiency?**

Ans. 35% to 75%.

Q. **Where the Tunger bulb rectifier is commonly used?**

Ans. In battery charging.

MERCURY ARC RECTIFIER

Q. **What is the type of mercury arc rectifier, whether the hot cathode or cold cathode?**

Ans. Cold cathode.

Q. **In a single phase rectifier how many main electrodes are there?**

Ans. Two.

Q. **Which is the cathode?**

Ans. Pool of mercury.

Q. **Is there any circuit to keep the mercury arc rectifier ready to deliver current?**

Ans. Yes, it is excitation circuit.

Q. **What is the main function of the excitation circuit?**

Ans. To start the rectifier and to maintain the hot spot even at no load.

Q. **What do you understand from the hot spot?**

Ans. In mercury arc rectifier, the heavy positive ions bombard the cathode surface and results a hot spot which emits the electrons.

Q. **How the positive heavy ions are produced and how hot spot is maintained?**

Ans. As the electrons, which are negatively charged particles, are released or emitted, these are attracted towards a positive charge i.e., anode plate. Now the electrons collide with the mercury vapours and ionize it (in positive and negative ions). The free electrons are attracted towards anode and positive heavy ions bombard on mercury pool causing an increase in temperature and maintaining a white spot on the surface of mercury pool.

Q. **The container is made of which material?**

Ans. Glass or steel. Glass bulb is used for low capacity and steel tank for heavy currents.

Q. **Can you tell the surface of mercury, whether concave or convex?**

Ans. Convex.

Q. **The glass bulb rectifiers are made up to which capacity?**

Ans. Upto 30 kW.

Q. **The steel tank rectifiers are used up to which capacity?**

Ans. 2500 kW.

Q. **How much maximum voltage can be rectified by the rectifier?**

Ans. Up to 20 kV, but with some special arrangement it can go upto 30 kV.

Q. **How much maximum current can be obtained?**

Ans. Up to 1600 amp.

Q. **How much is the voltage drop in the rectifier?**

Ans. 12 to 15 volts.

Q. **Why mercury is preferred for cathode?**

Ans. Because of its quality of self-restoring, high atomic weight, low specific heat, low boiling point, small latent heat of vaporisation and being its low ionization potential.

Q. **What are the advantages of mercury arc rectifier?**

Ans. It is simple in construction, easy in operation, occupies less space, light in weight, maintenance is less, starting time is less, high efficiency and overload capacity, no wear and tear and respond quickly to varying load demands etc.

Q. **On which voltage rating it gives better performance?**
Ans. From 200 V to the highest values of voltage required for commercial purpose.

Q. **How much overloading can last for how many hours for commercial purpose?**
Ans. 150% overloading for 2 hrs., 200% overloading for 5 minutes and 400% overloading for 15 seconds.

Q. **The traction purpose rectifiers are designed for how much capacity?**
Ans. 2 MW at 600 V.

Q. **The rectifiers in use are of how many phases?**
Ans. Single phase, three phase and six phase.

Q. **How many anodes are there in a three phase and six phase rectifier?**
Ans. Three and six, respectively.

Q. **How the cathode is energised?**
Ans. Separate winding is done for energising the cathode mounted on the same limbs of the transformer.

Q. **Is it possible to use the rectifier as an inverter?**
Ans. Yes, it can convert D.C. into A.C. by means of grid controlled mercury arc rectifier.

Q. **What are the different types of metal rectifier?**
Ans. Copper oxide, selenium and germanium rectifier.

Q. **In which direction the current flows in copper oxide rectifier?**
Ans. From oxide to copper.

Q. **If only one rectifier is in use, which type of rectification will be obtained?**
Ans. Half-wave rectification.

Q. **What is the use of obtaining full-wave rectification?**
Ans. To obtain a continuous and smooth D.C.

Q. **How and what arrangement is done to obtain full-wave rectification?**
Ans. Two rectifiers are connected or bridge rectifiers are used.

Q. **What is the safe reverse voltage for a rectifier?**
Ans. About 9 volts.

Q. **What is the average current rating for a disc of 3/4″ diameter?**
Ans. 1/3 amp.

Q. **What voltage can be applied to single rectifier?**
Ans. Ranging from 4 volts to 8 volts.

Q. **How 220 volts is rectified?**

Ans. By connecting a number of rectifiers in series say for 4 volt one unit and for 220 volts 55 units are connected in series.

Q. **For more current how these plates are arranged?**

Ans. In parallel.

Q. **What is the maximum current density in the forward direction for copper oxide rectifier?**

Ans. It is 0.1 to 0.15 A/cm^2 at an allowable voltage of 8 volts per element.

Q. **What is the thickness of the layer of film in case of selenium rectifier?**

Ans. 0.5 mm.

Q. **What is the reverse voltage for selenium rectifier?**

Ans. 13 volts but usually rated for 6 volts.

Q. **What is the maximum current density of selenium rectifier?**

Ans. 35 to 40 mA/cm^3.

Q. **Which rectifier has higher efficiency (i) copper oxide (ii) selenium?**

Ans. Copper oxide rectifier.

Q. **Where the metal rectifiers are used?**

Ans. Battery charging, electroplating, X-rays works, instrument operating relays, etc.

Q. **How many rectifiers are used in 3 phase full-wave rectification?**

Ans. Three.

Q. **How much output voltage is obtained in three phase?**

Ans. $0.83E_m$.

Q. **How many rectifiers are connected for six phase rectifier?**

Ans. Six.

Q. **How the secondary of the transformer is connected in case of three phase, six phase transformation and what is the polarity?**

Ans. In star and star point is negative.

OBJECTIVE TYPE MULTIPLE CHOICE QUESTIONS

1. Motor generator, motor converter and ...... are used to convert A.C. into D.C.

   (a) series generator (b) rotary converter
   (c) booster (d) induction motor.

2. The number of sliprings in three phase rotary converter are ......

   (a) two (b) three (c) four (d) six.

3. In case of single phase rotary converter, if it has 100% the ratio of voltage conversion is $E_{p3} = 0.707\ E_{dc}$. Here the current ratio is ……
   (a) $I_p = 0.414\ I_{dc}$ (b) $I_p = 1.414\ I_{dc}$
   (c) $I_p = 0.707\ I_{dc}$ (d) $I_p = 0.637\ I_{dc}$
4. The continuous bombardment of positive ions helps to maintain this high temperature of cathode which is known as ……
   (a) dry spot (b) hot spot
   (c) pool spot (d) black spot.
5. In mercury arc rectifier the anodes are made of ……
   (a) carbon (b) zinc (c) graphite (d) copper.
6. The surface of the mercury is ……
   (a) convex (b) concave (c) plane (d) parabolic.
7. The current of rectifier can be controlled with the help of perforated …… placed between the anode and the cathode
   (a) gas (b) grid
   (c) level of mercury (d) anodes.
8. In case of three phase rotary converter, if the voltage conversion is $E_p = 0.612\ E_{dc}$, then the current conversion between the two armature taps is ……
   (a) $I_p = 1.414\ I_{dc}$ (b) $I_p = 0.707\ I_{dc}$
   (c) $I_p = 0.943\ I_{dc}$ (d) $I_p = 0.544\ I_{dc}$
9. For conversion of three phase supply into six phase, there are three primaries and two similar secondary windings for each phase, the six phase windings can be connected in ……
   (a) double star only (b) double delta only
   (c) diametrically only
   (d) in double star or double delta or diametrical.
10. Which system of six phase secondary windings is commonly in practice ……
    (a) double star (b) double delta
    (c) diametrical (d) None of the above.
11. Which is the usual range of rotary converter ……
    (a) 1 kW to 1000 kW (b) 0 to 100 kW
    (c) 3 kW to 3000 kW (d) above 5000 kW.
12. The igniting of the mercury arc rectifier is commonly preferred by ……
    (a) by tilting the rectifier so that the displaced mercury comes into contact with the ignition electrode and then returning the rectifier to its normal position.

(b) by lowering and rising (vibrating) the ignition electrode on the mercury surface by means of a solenoid.

(c) by providing extra heating device for vaporisation of mercury.

(d) None of the above.

13. In case of mercury arc rectifier, more smooth D.C. rectification is obtained by using ......

(a) a single phase half-wave rectification.

(b) a single phase full-wave rectification.

(c) a three phase full-wave rectification.

(d) a six phase full-wave rectification.

14. In case of selenium type rectifier the steel disc is deposited with a ...... layer of selenium a hard grey coloured elements.

(a) thin (b) thick (c) normal (d) no.

15. The forward average current rating for a copper oxide rectifier of 20 mm dia, is about 0.33 amp., here the average maximum reverse safe voltage is ......

(a) 9 V (b) 20 V (c) 5 V (d) 10 V.

16. The copper losses (I^2R) in rotary converter are ...... times as compared to the D.C. generator of the same size.

(a) four (b) double (c) less (d) half

17. The positive ions in a mercury arc rectifier are attracted towards ......

(a) cathode (b) pool of mercury

(c) anode (d) auxiliary anode.

18. The form factor for a half-wave rectifier is ......

(a) 1.57 (b) 1.11 (c) 1.44 (d) 2.22

19. The r.m.s. value of a half-wave rectified sine wave current is ......

(a) $0.51 I_{max}$ (b) $0.637 I_{max}$ (c) $0.707 I_{max}$ (d) $1.44 I_{max}$

20. If the connections of anode and cathode of a mercury arc rectifier are interchanged then ......

(a) the rectifier will not function.

(b) the rectifier will work at 100% efficiency.

(c) the internal losses and drops will reduce.

(d) the white spot will be maintained at anode and not on cathode.

CHAPTER 22

Generation, Transmission and Distribution

INTRODUCTION

Q. **What do you mean by the generation?**

Ans. To generate electrical energy is known as generation.

Q. **What are the sources of generation?**

Ans. There are two types of sources—the primary sources like sun, wind, tide and secondary sources like coal, fuel, oil or gases, water and atomic fission, etc.

Q. **Which type of sources are commonly used in India?**

Ans. Secondary sources.

Q. **Name the type of power stations?**

Ans. Diesel power station, hydroelectric power station, thermal power station and nuclear power stations.

Q. **What is the name of the fuel generally used is diesel engine?**

Ans. Diesel.

Q. **What is the prime mover?**

Ans. The equipment which derives or gives mechanical power to the generator is called prime mover.

Q. **Name the generating station in which steam is used for driving the turbine?**

Ans. Steam or thermal power station.

Q. **Name the turbine which is used for steam power station?**

Ans. Steam turbine.

Q. **Name the different stages of steam power station?**

Ans.
1. Coal and ash handling arrangement.
2. Steam generating plant.

3. Steam turbine.
4. Electrical plant.
5. Feed water and cooling arrangement.

Q. **What important considerations are to be kept in mind for selecting the site for steam power plant?**

Ans. The site selected should be near the coal and water or say coal and water should be available in plenty. Power accessibility and extension of the plant side etc., should be kept in mind.

Q. **What is the function of cooling system?**

Ans.
1. It removes the heat energy of the exhaust steam.
2. It keeps the turbine back pressure at the lowest possible level.

Q. **Which power plant has less production cost?**
(i) Steam power plant (ii) Diesel power plant.

Ans. Steam power plant.

Q. **What is the main disadvantages of thermal power stations?**

Ans. It pollutes the atmosphere due to smoke and dust particles etc.

Q. **Can you tell the overall efficiency of thermal power station?**

Ans. About 25%.

Q. **Which is called the hydroelectric power station?**

Ans. The power stations which convert the kinetic energy of water into electrical energy, are called hydroelectric power stations.

Q. **What is the prime source of hydroelectric power station?**

Ans. Water in large quantity.

Q. **Can you name the place where it is stored?**

Ans. Water reservoir.

Q. **What is the dam in hydroelectric power stations?**

Ans. It is a masonry construction and structure which is used to create water head and store water in catchment area.

Q. **Name the huge pipe used for bringing the water to the turbine?**

Ans. Pen stock.

Q. **What is the water turbine?**

Ans. It is used to give mechanical energy to the alternator. It is driven by the water flow or the energy stored by water.

Q. **What are the types of the turbine?**

Ans. Generally two—the impulse turbine and reaction turbine.

Q. **What is the unique advantage of hydroelectric power station?**

Ans. It helps the irrigation.

Q. **What is the disadvantages of the power station?**

Ans. The initial cost is high, huge quantity of water is required.

Q. **Where the hydroelectric power stations are installed in India?**

Ans. Bhakra Nangal, Chambal project, Rana Partap Sagar dam and Koyna etc.

Q. **What is the nuclear power station?**

Ans. The electric power generating stations in which the nuclear energy is converted into electrical energy are known as nuclear power stations.

Q. **Name the fuel generally used for nuclear power stations?**

Ans. U^{235} or uranium, Tr^{232} or thorium.

Q. **How heat is released?**

Ans. By fission.

Q. **How the steam in produced?**

Ans. The heat produced by fission is used to boil water or producing steam which drives the turbine.

Q. **Name the different stages of nuclear power stations?**

Ans. Nuclear reactor, heat exchanger, steam turbine and alternator.

Q. **What is the reactor?**

Ans. In nuclear reactor the uranium, U^{235} is subjected to nuclear fission or the chamber where this fission takes place is known as nuclear reactor.

Q. **How much kWh energy is released by the splitting of one kg of uranium U^{235}?**

Ans. One kg of uranium yields 25×10^6 kWh in heat form.

Q. **How much coal energy is equivalent to one kg of uranium for heating purpose?**

Ans. The energy obtained by burning about 2500 tonnes of high grade coal.

TRANSMISSION

Q. **What is meant by the transmission of electrical energy?**

Ans. To deliver bulk power from power station to the load centre at higher potential is known as transmission.

Q. **What is voltage range for transmission purpose?**

Ans. The voltage is high for long transmission. It is about 110 kV, 132 kV, 220 kV and 440 kV. For short transmission it is 11 kV, 33 kV and 66 kV etc.

Q. **Give the reason, why we are using high voltage for transmission?**

Ans. Size of the conductor is reduced as the current is reduced less copper required, less power loss and drops, efficiency is good, maintenance cost is less, number of supports are less.

Q. **What are the different methods of transmission?**

Ans. Overhead transmission and underground transmission.

Q. **Which system is safe, the UG system or OH system?**

Ans. UG system.

Q. **Which system is costly?**

Ans. UG system.

Q. **Generally which system of distribution is used for high voltage transmission and why?**

Ans. Overhead due to low cost.

Q. **In which system the repairing is easy?**

Ans. Overhead system.

Q. **What do you know about the O.H. system?**

Ans. In this system the conductors are run sufficiently high above the ground level supported by poles and insulators?

Q. **Name the conductors generally used for O.H. line?**

Ans. Generally (i) Hard drawn bare copper conductor (ii) Aluminium cored steel reinforced (ACSR) conductor.

Q. **Can you use G.I. wire for O.H. line?**

Ans. Yes, it can be used for short distance distribution lines of O.H. system.

Q. **What is underground system?**

Ans. In this system the cable runs under the ground.

Q. **What properties are required for an insulation?**

Ans. Should have good mechanical strength, good dielectric strength, insulation should be moisture proof and should be free from mechanical damages.

Q. **Which system has longer life?**

Ans. U.G. system.

Q. **In which system the extension is easy?**

Ans. O.H. system.

Q. **Can you locate fault in U.G. system easily?**

Ans. No, the fault location is easy in case of O.H. system.

DISTRIBUTION

Q. **What is the distribution system?**

Ans. To distribute electrical energy to the consumers for use is known as the distribution.

Q. **What is A.C. distribution?**

Ans. If A.C. energy is being distributed then it is called as A.C. distribution.

Q. **What do you call the system in which D.C. supply is being distributed?**

Ans. D.C. distribution.

Q. **What is the substation?**

Ans. Substation serves as the source of energy supply for local areas of distribution in which these are located.

Q. **What are the classifications of the substations?**

Ans. These are :

1. Transformer substation.
2. Converting substation.
3. Frequency substation.

Q. **What are the classifications of the substation according to the design?**

Ans.

1. Indoor substation.
2. Outdoor substation.

The outdoor substations are designed :

(a) Pole mounted.

(b) Foundation mounted.

Q. **What is a transformer substation?**

Ans. In this type of substation, the voltage is either step up or step down by means of the transformer.

Q. **What do you know about the converting substation?**

Ans. In this type of substation, the A.C. energy is converted into D.C. energy.

Q. **Can you name the substation, in which the frequency is changing from one value to the another value with the help of frequency converter?**

Ans. Frequency substation.

Q. **Can you tell the types of distribution according to the voltage?**

Ans.

1. High voltage distribution.
2. Low voltage distribution.

Q. **What are the main methods for D.C. distribution?**

Ans.

1. Two wire distribution.
2. Three wire distribution.

Q. **Can the coding of the wire be known in three wire system?**

Ans. Yes, it is positive outer, negative outer, and middle wire.

Q. **How three wire distribution is obtained?**

Ans.

1. By using two generators.
2. By using one generator of more voltage.

Q. **What are the main accessories and equipments used for substation?**

Ans. Main busbar, insulator, isolator, circuit breaker, power transformer, indicating and measuring instruments, instrument transformers, fuses, protective relays and control cable and circuit etc.

Q. **What are the main functions of circuit breaker?**

Ans.
1. It carries current continuously without heating.
2. Opens or closes the circuit when required.
3. It protects the circuit in case of any fault (short circuit, earth fault, on overloading etc.).

Q. **How the contacts of circuit breaker are classified?**

Ans. The moving contacts, the fixed contacts.

Q. **Give some names of circuit breakers?**

Ans. Air circuit breaker, SF_6 circuit breaker, oil circuit breaker, water type circuit breaker, air blast circuit breaker etc.

Q. **What do you mean by a circuit breaker?**

Ans. A circuit breaker is essentially a switching and circuit interrupting device.

Q. **What are the main purposes of a circuit breaker.**

Ans. It serves dual purpose, the switching on and switching off in normal conditions of operations and secondly protecting the entire circuit during abnormal (short circuit, earth fault, etc.) conditions.

Q. **Precisely a circuit breaker does which type of functions?**

Ans. The functions precisely can be classified in two types—the arc extinction and the relay operations.

Q. **Why arc extinction is so important?**

Ans. Whenever a is circuit is on and current is interrupted by the circuit breaker, arc formation is inherent between the contacts which does not cease the current to zero ampere but it delays the decay for some milliseconds. The arc has a great deal of temperature say few thousand degrees which is sufficiency to cause any damage to the contacts and circuit too.

Q. **Can you give some idea about an arc?**

Ans. It is consisting of a column of ionised gases.

Q. **Can you tell whether the ionisation results only light?**

Ans. No, this ionisation of gases is accompanied by the emission of light and heat.

Q. **Can you give some elementary idea of the arc temperature so developed?**

Ans. It may be as high as 5000°C or so.

Q. **Which methods of arc extinction are generally adopted?**

Ans. High resistance interruption and low resistance interruption.

Q. **What is the main aim of high resistance interruption procedures?**

Ans. It is to reduce the current to too small as to be insufficient to maintain the ionisation process.

Q. **What is low resistance interruption procedure?**

Ans. It is to lower the arc resistance to be low to maintain the arc energy to a minimum value. The use of natural or artificial current zero and then prevented from restriking.

Q. **What is arc dispersion?**

Ans. In this process the destructive phenomenon of arc is eliminated/minimised.

Q. **What are the methods adopted for that dispersion?**

Ans. Oil immersion of contacts, multiple break contacts, de-ionization of arc path, magnetic blow out of arc, and blast principle using air, oil, gas or water.

Q. **Is there any process in which recurrence or arc is prevented by reducing current and voltage below the minimum arcing value?**

Ans. Yes, it is a arc prevention process.

Q. **How is this achieved?**

Ans. By discharge resistance, rectifiers, and condensers.

Q. **Do you have any idea of the media used in air circuit breaker, air blast circuit breaker and high voltage direct current CB?**

Ans. Yes, in air breakers the air at atmospheric pressure, in air blast circuit breaker compressed air between 20 to 30 atmospheric pressure, and for HVDC CB the vacuum or SF_6 as medium.

Q. **What types of relays are generally used in circuit breakers?**

Ans. Differential relay, impedance relay, over-current relay, instantaneous relay, static relay.

Q. **What is the function of differential relay?**

Ans. These relays responds to vector difference between two or more similar electrical quantities.

Q. **Name the relay which operates when the impedance between relay point and fault point is below a specified value.**

Ans. Impedance relay.

Q. **What do you call a operating relay which operates in less than 0.2 second operation time?**

Ans. It is known as the instantaneous relay.

Q. **What is the function of overload relays?**

Ans. These relays function in abnormal conditions of overloading.

Q. **Is there any static relay? What is the function of this relay?**

Ans. It is a relay used in circuit breakers. It is a relay without moving parts. In this relay the measurement is performed by a stationary circuit.

Q. **What is arcing time?**

Ans. It is the time between separation of the circuit breaker contacts and final current to be zero.

Q. **Are we using any 'dead time' term for the circuit breaker?**

Ans. It is the measure of time between final current zero ampere of first opening and contact touch during subsequent reclosing.

Q. **Essentially speaking, is the circuit breaker, a current or voltage interrupting device?**

Ans. It is a current interrupting device.

Q. **What is the category given to the circuit breakers having voltage rating 1000 V?**

Ans. Low voltage circuit breakers.

Q. **The fault clearance time of circuit breaker is usually given in few hours or few minutes or few cycles of A.C. supply voltage?**

Ans. Few cycles of supply voltage.

Q. **Are you aware of the air pressure in the air blast circuit breakers?**

Ans. In general for ABCB the air pressure ranges from 20 to 30 kg/cm^2.

Q. **Nowadays we are switching on to SF_6 circuit breakers, can you give any idea about this gas circuit breaker?**

Ans. The SF_6 is the sulphur hexafluoride gas. This gas is highly non-inflammable, the density being five times more than the air. The pressure used for these circuit breakers ranges from 3 to 5 kg/cm^2.

Q. **Name the properties to be considered while selecting a gas medium for a circuit breaker.**

Ans. The high dielectric strength, non-inflammability, non-toxicity.

Q. **The insulation resistance of a high voltage circuit breaker should be approximately how much?**

Ans. About 2000 MΩ.

Q. **The active contacts of a circuit breaker are made up of which materials?**

Ans. These are made of copper tungsten alloy.

Q. **The air circuit breaker has SF_6 as the arc quenching medium?**

Ans. No, it has only air as arc quenching medium.

Q. **The air blast circuit breakers can stand up to approximately which capacity of loadings?**

Ans. It can withstand up to so many kA.

Q. **Can you give any idea about the normal capacity of fuse beyond which it is not used?**

Ans. Approximately 100 A, beyond this it is not a good protection system.

Q. **Are you having any idea about the operating time for which ABCB are designed to operate on 400 kV power system.**

Ans. It is approximately 50 milliseconds.

Q. **Define the protective relay?**

Ans. It may be defined as an electrical device interposed between the main circuit and the circuit breaker in such a manner that any abnormality in the circuits acts on the relay which in turn in case of faulty situation causes the circuit breaker to isolate the circuit.

Q. **A relay is supposed of how many important elements?**

Ans. It can be of three different elements namely sensing element, comparing element and control element.

Q. **Can you give any other name to the sensing element?**

Ans. Yes, it is also known as measuring element.

Q. **The distribution line is subdivided into some elements, name them?**

Ans. Feeders, distributors and service mains.

Q. **What is a feeder?**

Ans. A feeder is said to be the conductor which connects the station, generally generating station, to the areas to be fed by those stations.

Q. **Can you connect load on the feeders or can you take any tapping from the feeders?**

Ans. No.

Q. **What do you known about distribution?**

Ans. To distribute the electrical energy to the consumers is known as distribution, and the conductor from which a number of tappings are taken is known as distributor.

Q. **Give some idea about service mains?**

Ans. Service mains are those which connect the consumer's terminals to the distributors.

Q. **Why A.C.S.R. conductors are used for transmission line?**

Ans. To increase the transmission efficiency. For minimising the corona effect effectively and to increase the mechanical strength of the conductors the steel wire is used in the central core of the cable.

Q. **What should be the properties of good conductor?**

Ans. Low specific resistance, good mechanical strength, appreciable ductility, easily jointable and free from the atmospheric changes.

Q. **Which system of transmission and distribution will you prefer for a thickly populated area?**

Ans. U.G. system (underground system).

Q. **What is a balancer?**

Ans. It is a generator which helps in keeping the potential difference of the middle wire a balanced one. These are the motor generator set mechanically coupled.

Q. **What are the different type of balancers?**

Ans. Rotary and static balancers.

Q. **What is a booster?**

Ans. It is a series generator the voltage of which is added to line to compensate voltage drop in the line.

Q. **Can you name the booster which is connected in feeders?**

Ans. Feeder booster.

Q. **What are the main materials used for overhead line?**

Ans. The conductor, supports, insulators, cross arm etc.

Q. **What is the span?**

Ans. It is the horizontal distance between two adjacent poles.

Q. **What do you understand from the ground clearance?**

Ans. It is the distance between ground and the minimum level of the line conductor.

Q. **On what factor the ground clearance depends?**

Ans. Voltage.

Q. **What do you call the distance between the line conductors?**

Ans. Line clearance.

Q. **What is sag?**

Ans. It is the distance between the lowest and highest level of the wires in one span.

Q. **Which factors affect the sag?**

Ans. The span, weight of line, wind, thermal stresses of the atmosphere etc.

Q. **What do you call the depth of the pole which is buried under the ground?**

Ans. Setting depth.

Q. **Name the poles generally used for distribution purposes?**

Ans. Wooden poles, iron poles (steel poles), concrete poles and sometime towers.

Q. **What do you know about cross arm?**

Ans. These are prepared from either steel or wood and are fixed on the top leaving some distance. The insulators etc. are fixed on them. The cross arm and pole are perpendicular to each other.

Q. **Where the wooden poles are used?**

Ans. Generally in hilly areas where wood is available in plenty.

Q. **What are steel poles and what are their classifications?**

Ans. These poles are prepared from steel. They are rail poles, tubular poles, and sometimes fabricated poles.

Q. **Can you tell the approximate setting depth of the poles?**

Ans. Approximately 1/5th of the pole length.

Q. **What do you known about tubular poles?**

Ans. The poles which have tubular construction. These may have or may not have the radius in steps or gradual steps, for example, the telephone poles and street light poles etc.

Q. **How the rail poles are prepared?**

Ans. From the rails.

Q. **Where concrete poles are used?**

Ans. These are used for distribution and transmission purposes. There the iron bars are embedded so that earth continuity is maintained.

Q. **How is the life and appearance of the concrete pole?**

Ans. It is having long life and better appearance.

Q. **What are the towers?**

Ans. These are used for long transmission and high voltage. These are fabricated from angle iron and steel sections etc.

Q. **What are the types of towers?**

Ans. Rigid and flexible.

Q. **Which are right and flexible towers?**

Ans. The towers which have four or three legs are known as rigid towers but the tower which has only two legs is known as flexible tower.

Q. **How the wires are supported on the cross arm?**

Ans. With the help of insulators.

Q. **What are the types of insulators?**

Ans. Pin insulator, shackle insulator, suspension type insulator, egg or stay insulator.

Q. **How pin insulators are supported on the arms?**

Ans. By means of pin and screws.

Q. **Which material is used for insulator?**

Ans. The material is usually porcelain, glass stealite and sometimes special composition materials are also used.

Q. **What is the dielectric strength used for the insulator made of porcelain?**

Ans. Approximately 60 kV/cm of its thickness.

Q. **Upto which voltage the pin insulators are used.**

Ans. Upto 33 kV but can be used upto 50 kV.

Q. **What is shackle insulator and where these are used?**

Ans. These shackle insulators are used to support wires at one side where the wires take a bend at a particular angle. These are supported on the screws in the centre with nuts and bolts.

Q. **What are the disc insulators or suspension type insulators?**

Ans. These insulators are used for high voltage transmission. The wires are supported by three insulators and these are generally in suspended position.

Q. **What is the dielectric strength of these insulators?**

Ans. Generally these are designed for 11 kV each and generally used for 33000 volts line and above.

Q. **Where the shackle insulators are used?**

Ans. At dead ends or at corners.

Q. **What is egg insulators and where these are used?**

Ans. These insulators are of egg shape and are used in stay wire section to avoid any leakage to earth and also to avoid any danger of shock etc.

Q. **What is stay wire?**

Ans. These are generally galvanised steel wires used to avoid bending of pole due to tension of overhead lines.

Q. **What are the advantages of guarding on O.H. line?**

Ans. To save the life, lines and any other type of accident due to electricity.

Q. **How guarding is helpful?**

Ans. As soon as the wire breaks, and if it is energised, it may be dangerous. But before coming to ground, it will touch the guard wire which is earthed, thus this short circuit will trip the circuit breaker in the sub-station and isolate the line. Thus the life and line both are saved.

Q. **Which type of guarding is used for railway crossing?**

Ans. Cage type.

Q. **Which type of guarding is done on road crossing?**

Ans. Cradiel type.

Q. **What precautions should be taken for erecting the overhead line?**

Ans. All parts of O.H. line except the current-carrying conductors must be earthed. The earthing must be done at four points in a mile.

Q. **Name the insulating materials used for underground cables?**

Ans. Oil impregnated paper, natural or synthetic vulcanised rubber compound, polythene and polyvinyl chloride, gutta-percha, silk or cotton and enamel insulations.

Q. **What are the types of supertension cables?**

Ans. H type cables, SL type cable, HSL type cable.

Q. **Name the extra high tension cables?**

Ans. Oil filled cable, gas pressure cable (internal pressure, external pressure cables).

LIGHTNING ARRESTER

Q. **What for the lightning arrester is used?**

Ans. It is used for the safety from the sky lightning or a surge voltage.

Q. **Where it is installed?**

Ans. On the top of the building, top of the poles and towers.

Q. **What are the different types of lightning arresters?**

Ans. Rod gap arrester, horn gap arrester, electrolytic arrester, oxide film arrester, thyrite arrester, expulsion type lightning arrester and valve type arrester etc.

Q. **What is the full form of PILC, PILCSTA and PILCDTA cables?**

Ans. PILC—Paper insulated lead covered cable.
PILCSTA—Paper insulated lead covered single taped armoured cable.
PILCDTA—Paper insulated lead covered double taped armoured cable.

OBJECTIVE TYPE MULTIPLE CHOICE QUESTIONS

1. The most common generation voltages in India are ......
   (a) 440 V, 6.6 kV, 11 kV (b) 440 V, 220 V, 110 V
   (c) 11 kV, 22 kV, 66 kV (d) 11 kV, 13.2 kV, 18 kV.
2. The most common transmission voltage in India are ......
   (a) 33 kV, 132 kV, 20 kV (b) 132 kV, 220 kV, 440 kV
   (c) 220 kV, 440 kV, 500 kV (d) 11 kV, 132 kV, 500 kV
3. The towers are of two types ......
   (a) flexible and rigid (b) long and short
   (c) thick and thin (d) copper and aluminium
4. In thickly populated area the distribution lines are drawn using ......
   (a) underground system (b) overhead system
   (c) trench system (d) towers.
5. The service main drawn from pole to the meter is the property of ......

(a) consumer
(b) supplier
(c) consumer and supplier
(d) None of the above a, b, c.

6. The guard wire is firmly connected with ......
   (a) neutral wire
   (b) phase wire.
   (c) cross arm
   (d) earth wire.
7. The ...... insulators are used for the E.H.T. transmission line ......
   (a) pin
   (b) shackle
   (c) suspension
   (d) strays.
8. The load must not be connected directly on ......
   (a) transmission line
   (b) distributor
   (c) service mains
   (d) L.T. 220 volt line.
9. The lightning arrester is used ......
   (a) for the safety of the transformer
   (b) for the safety of the line
   (c) for the safety of the consumer
   (d) for the safety of line, connected instruments, apparatus, machines, and the consumer also.
10. The lightning arrester is installed ......
    (a) at the top of the building
    (b) at the bottom of the building
    (c) in the centre of the building
    (d) away from the building to be protected.
11. The ...... insulators are used on 132 kV transmission line.
    (a) shackle type
    (b) pin type
    (c) disc or suspension type
    (d) egg type.
12. Generally the porcelain insulators are glazed because ......
    (a) it helps in gliding down of water easily
    (b) it improves its appearance
    (c) it prevents the absorbing tendency of gases and water vapours
    (d) it increases its mechanical strength.
13. In case of pin type insulators, the petticoats are provided to ......
    (a) improve its appearance
    (b) to drip out the rain water and to increase the possible conducting length of the path
    (c) to protect the stud from sun lighting.
    (d) to protect the stud from wind pressure.
14. The skin effect ...... with the ...... of frequency of supply system.
    (a) increases; increasing
    (b) decreases; increasing

(c) decreases; decreasing (d) increases; decreasing

15. The transformer at the sending end of the transmission line is connected in ......
 (a) star-star (b) star-delta (c) delta-star (d) delta-delta.

16. State the following statements, whether true or false.
 (i) Nuclear energy is not an unconventional source of energy.
 (ii) Coal broken into fine particles is known as pulverised coal.
 (iii) The coal generally used in thermal power station has the heating value approximately 5000 to 6500 kcal/kg.
 (iv) Heating value of coal largely depends upon the size of coal particles.
 (v) Bituminous coal is a superior quality of coal.
 (vi) The storage of coal which is sufficient to meet 24 hours demand of the plant is known as 'live storage'.
 (vii) Ball mill is used to pulverise the coal.
 (viii) In India, Bharat Heavy Electricals Ltd. is manufacturing thermal power plants.
 (ix) Bhakra Nangal Dam is constructed on Beas.
 (x) Koyna dam is in Maharashtra.

17. Which of the following cities is having atomic power station.
 (a) Delhi (b) Trombay
 (c) Yamuna Nagar (d) Agra.

18. Solar cells are made of ......
 (a) copper (b) aluminium (c) silicon (d) silver.

19. A single solar cell has ......
 (a) 0.5 V (b) 0.2 V (c) 1.5 V (d) 2 V.

20. The solar cell output is approximately ......
 (a) 0.5 W (b) 1 W (c) 2 W (d) 0.2 W.

21. The generation of power by solar cell has great disadvantage that the power generation is ......
 (a) very costly (b) very cheap
 (c) requires more space (d) requires more labour.

22. In satellite, source of energy is ......
 (a) thermal power system (b) battery
 (c) solar cells (d) None of the above.

23. In India, the highest transmitting voltage is ......
 (a) 400 kV (b) 11 kV (c) 132 kV (d) 7500 kV.

24. Up to which voltage the pin insulators are preferred?
 (a) 25 kV (b) 11 kV (c) 500 kV (d) 1000 V.

25. Name the effect which is resulted because of the size of conductor, frequency of current and the specific resistance of the conducting material.
    (a) Corona effect (b) Skin effect
    (c) Thomson effect (d) None of the above.

26. If the specific resistance of the conductor is increased, what effect does it result on the skin effect?
    (a) Increasing (b) Decreasing
    (c) Remains the same (d) None of the above.

27. In which season the load on the generating power station will be highest?
    (a) Summer (b) Winter
    (c) Rainy (d) None of the above.

28. The capacitor bank is installed to ...... power factor.
    (a) improve (b) decrease
    (c) keep the same (d) neutralise.

29. The disadvantages of transmission line as compared to cable system are ......
    (a) atmospheric hazards
    (b) atmospheric hazards, exposures to lightening
    (c) interference of telecommunication circuits
    (d) atmospheric hazards like smoke, ice, fumes, exposures to lightening, interference of telecommunication circuits, etc.

30. Which distribution system is more reliable?
    (a) Ring main system (b) Tree system
    (c) Two wire system (d) Three wire system

31. While the busbars are rated, which factor is taken into account?
    (a) Current and voltage
    (b) Voltage and frequency
    (c) Current, voltage, frequency and short time current
    (d) Short time current and frequency.

32. The sulphur hexafluoride (SF_6) is a gas which is used as a interrupting medium in a contractor of circuit breaker, it is used in the CBs at a pressure ranging from ......
    (a) 1 to 2 kg/cm^2 (b) 3 to 5 kg/cm^2
    (c) 7 to 10 kg/cm^2 (d) 10 to 15 kg/cm^2

33. The insulation resistance of the H.T. CBs in no case be less than ......
    (a) 2000 MΩ (b) 1 MΩ (c) 1000 MΩ (d) 5000 MΩ

34. The safe working life of the underground cable is taken as ......
    (a) 2 years (b) 5 years (c) 40 years (d) 100 years.

35. Pole mounting substations are also known as ...... substations.
    (a) outdoor type
    (b) indoor type
    (c) basement type
    (d) open type
36. A circuit breaker is a ...... and ...... device.
    (a) switching; protecting
    (b) switching; measuring
    (c) interrupting; measuring
    (d) None of the above.
37. A circuit breaker serves the purpose of ......
    (a) switching on and switching off under normal working conditions.
    (b) protecting the circuit under abnormal conditions of load.
    (c) protecting the circuit under faulty conditions i.e., short circuit, earth fault, overloading etc.
    (d) All the above.
38. If the current is flowing/jumping through ...... it is called an arc.
    (a) air
    (b) solids
    (c) conductors
    (d) semiconductors.

CHAPTER **23**

Electronics

INTRODUCTION

Q. **What do you understand by electronics?**

Ans. The branch of engineering and science which deals with the flow of electrons or where the conduction of electrons takes place through gas, semi-conductor or vacuum.

Q. **The flow of electrons constitutes which effect?**

Ans. Electric current.

Q. **What is the mass, charge and radius of electron?**

Ans. Mass = 9×10^{-31} kg; charge = 1.602×10^{-19} coul; radius = 1.9×10^{-15} metre.

Q. **Can you tell the methods of electron emission?**

Ans. Thermionic emission, photoelectric emission, field emission, secondary emission and radioactive disintegration.

Q. **What is thermionic emission?**

Ans. The phenomenon in which the electrons are emitted from the metal surface by supplying the thermal energy to it is known as thermionic emission.

Q. **Name the type of emission in which high velocity beam of electrons or ions strikes the metal surface and electrons are released?**

Ans. Secondary emission.

Q. **If on an emitter the light falls and liberates free electrons what do you call this phenomenon?**

Ans. Photoelectric emission.

Q. **In the field emission is the cathode heated up or remains cold?**

Ans. Cold.

Q. **How the field emission takes place?**

Ans. By applying a high positive voltage electric field.

Q. **Does the free electron in normal state moves or remains standstill?**

Ans. Moves.

Q. **If these move why these do not come out and current flows through the substance without applying the electric pressure?**

Ans. There is a energy barrier and the energy acquired by these electrons is not sufficient to come out, so it requires extra energy which is given by means of applying the electric pressure from outside.

Q. **What are the methods of heating the cathode?**

Ans. Direct heating and indirect heating.

Q. **Can you give further classifications of heating the cathode?**

Ans. Direct heating :

(i) Filament type

(ii) Heater cathode type.

Indirect heating : These are three :

(i) Pure tungsten filament

(ii) Throiated tungsten filament

(iii) Oxide coated filament.

Q. **Which type of heating is preferred nowadays?**

Ans. Oxide coated filament type.

Q. **What is a diode?**

Ans. It has two elements—a cathode and an anode. The cathode is an emitter of electrons and a plate surrounding the cathode acts as a collector of electrons.

Q. **What is the use of diode?**

Ans. As a half-wave rectifier?

Q. **Can you use the diode for full-wave rectifier?**

Ans. Yes, by using two diodes or four as a bridge rectifier.

Q. **What material is used for plate?**

Ans. Hollow metallic cylinder made of nickel, graphite tungsten, tantalum or iron etc.

Q. **What is a space charge?**

Ans. The dense cloud of electrons surrounding the cathode is called space charge.

Q. **When the anode is at negative potential with reference to cathode how much current conduction takes place?**

Ans. No current.

Q. **Can a control grid be kept at positive potential with respect to cathode?**

Ans. Never in use.

Q. **What is a rectifier?**

Ans. The electronic device which converts A.C. into D.C. is called rectifier.

Q. **What is a crystal diode?**

Ans. The PN or NP junction is known as semi-conductor or crystal diode.

Q. **What are the uses of triode?**

Ans. It can be used as an amplifier, oscillator etc.

Q. **How much time is the ripple frequency of the A.C. to supply frequency?**

Ans. Twice as the A.C. supply input for 50 c/s it is 100 c/s.

Q. **What is the ripple factor?**

Ans. It is the ratio of ripple voltage to A.C. voltage or ratio of r.m.s. value of A.C. component to average D.C. voltage of the wave.

Q. **What is the peak inverse voltage (PIV)?**

Ans. It is the maximum reverse voltage that occurs across the diode.

OBJECTIVE TYPE MULTIPLE CHOICE QUESTIONS

1. The thermionic values helps in the study of ......
   (a) atomic structure (b) atomic explosions
   (c) behaviour of electrons
   (d) internal construction of nucleus.
2. The emission in which liberation of electrons is by heat is known as ......
   (a) photoelectric emission (b) field emission
   (c) thermionic emission (d) secondary emission.
3. The minimum energy required to release electron from the surface of a metal is called ......
   (a) work function (b) dead function
   (c) exciting function (d) limiting function.
4. In a diode tube the electrons are released by ......
   (a) grid (b) heater (c) cathode (d) anode.
5. The filament of a thermionic diode can be heated ......
   (a) directly only (b) indirectly only
   (c) directly or indirectly (d) None of these.
6. The cloud of emitted electrons around the cathode is known as ......
   (a) space cloud (b) space charge
   (c) electron charge (d) neutron charge.
7. The ratio of emission current in amperes to the heating power in watts is known as ......

(a) power factor
(b) cathode emission efficiency
(c) anode emission efficiency
(d) field emission efficiency.

8. When the plate of a diode valve is more negative w.r.t. cathode the current flow is from ......
   (a) cathode to heater
   (b) cathode to plate
   (c) heater to cathode
   (d) stopped.
9. In a diode the current flows from ......
   (a) plate to cathode
   (b) cathode to plate
   (c) grid to plate
   (d) cathode to grid.
10. Generally the diode is used as ......
    (a) an amplifier
    (b) current inverter
    (c) a rectifier
    (d) voltage inverter.
11. The performance of a triode depends on three main factors ......
    (a) anode resistance, transconductance and amplification factors
    (b) anode, cathode and grid
    (c) anode, cathode and heater
    (d) cathode, resistance and amplification factor.
12. The 'bias' term is used in a triode refers ......
    (a) a reference plate voltage
    (b) the potential difference between plate and heater
    (c)
    (d) A.C. voltage applied to grid.
13. The most common cause of failure in a diode tube is ......
    (a) burned out filament
    (b) open circuit in grid
    (c) burned out grid
    (d) warped plate.
14. The ratio of the ΔI_p and ΔV_g is known as ......
    (a) transconductance
    (b) secondary conductance
    (c) primary conductance
    (d) bombardment
15. The ratio of ΔV_p and ΔV_g is known as ......
    (a) amplification factor
    (b) thermionic emission
    (c) rectification factor
    (d) peak factor
16. The tetrode is preferred instead of triode because of ......
    (a) amplification
    (b) rectification
    (c) higher frequencies
    (d) secondary emission.
17. When the ions are liberated because of the knocking of high speed electrons to plate, the emission is known as ......
    (a) thermionic emission
    (b) field emission
    (c) photoelectric emission
    (d) secondary emission.

18. The voltage amplification by a single triode is small to increase it, the output of one triode is given to the grid of second and the output of second to third triode grid, this phenomenon is known as ......
   (a) single stage amplification (b) two stage amplification
   (c) multistage amplification (d) None of these.

19. The valve which has got third grid or suppressor grid between the screen grid and plate is known as ......
   (a) diode (b) tetrode (c) pentode (d) triode.

20. A semi-conductor is one which ......
   (a) is a good conductor of electricity
   (b) is a good insulator of electricity
   (c) neither a conductor nor an insulator
   (d) carry an electric current uninterrupted.

21. Generally all the semiconductors are ......
   (a) monovalent (b) covalent
   (c) neighter mono- nor covalent (b) None of these a, b, c.

22. The N type semiconductors are made by adding impurity as ......
   (a) phosphorus, antimony and arsenic
   (b) aluminium, boron or selenium
   (c) copper, aluminium and carbon
   (d) copper, boron and arsenic.

23. The p-type semiconductors are made by adding impurity as ......
   (a) aluminium, antimony or arsenic
   (b) aluminium, boron or iridium
   (c) antimony boron or selenium
   (d) copper, carbon and selenium.

24. The majority carriers in a semiconductor material are ......
   (a) free electrons in P and N type semiconductor
   (b) holes in P and N type semi-conductor
   (c) free electrons in N type and holes in P type semi-conductor
   (d) holes in N type and free electrons in P type semi-conductors.

25. Name the low pass filter from the following ......
   (a) L-type with series L and shunt C
   (b) L-type with series C and shunt L
   (c) T-type with series L and shunt C
   (d) T-type with series C and shunt L.

26. The capacitor filters are preferred because it gives ......
   (a) large ripple with high voltage
   (b) large ripple with low voltage

(c) small ripple without voltage
(d) small ripple with high voltage

27. In case of indirectly heated cathode, the heater voltage ......
    (a) is applied to cathode
    (b) is separate from the cathode circuit
    (c) should be a variable D.C. voltage
    (d) is equal to the C-bias voltage.

28. The three phase rectification is preferred because ......
    (a) each secondary winding carries current for 1/3 of a cycle
    (b) each primary winding carries current for 2/3 of a cycle
    (c) Cu losses are comparatively less
    (d) each primary winding carries current for 2/3rd of a cycle, each secondary winding carries current for 1/3rd of a cycle, and the copper losses are comparatively low.

29. The harmful effect of secondary emission in a tetrode is eliminated by using an additional third grid between screen grid and plate, the valve is known as ......
    (a) pentode (b) octode
    (c) triode (d) None of the above.

30. The peak inverse voltage rating per diode, in case of bridge rectifier, is ...... as compared to full-wave rectifier.
    (a) half (b) twice (c) same (d) triple

31. The ripple frequency of a full-wave rectifier is ...... times the A.C. supply frequency.
    (a) 0.5 (b) 2 (c) 1.5 (d) 4.

32. In case of NPN transistor, negative voltage is needed at the collector.
    (a) True (b) False.

33. In an NPN transistor, positive voltage at the base w.r.t. emitter provides forward bias.
    (a) True (b) False.

34. Bridge rectifiers are preferred because they require small transformer and less PIV.
    (a) True (b) False

35. The effect of bypass capacitor in amplifier design is adversely seen on ...... frequency response.
    (a) low (b) high
    (c) medium (d) None of the above.

CHAPTER 24

Winding

INTRODUCTION

Q. **Which winding is known as closed type winding?**

Ans. D.C. armature winding.

Q. **A.C. winding is generally termed as which type of winding?**

Ans. Open winding.

Q. **Name the accessory where the leads of the armature are brought and soldered?**

Ans. Commutator.

Q. **How many types of D.C. winding is there?**

Ans. Lap winding and wave winding.

Q. **What do you mean by lap winding and wave winding?**

Ans. The winding in which the number of parallel paths are equal to the number of poles, that winding is called lap winding. The winding which has got only two parallel paths, whatsoever the number of poles may be is called wave winding.

Q. **Which winding is done for (i) high current (ii) high voltage?**

Ans. (i) Lap winding.
(ii) Wave winding.

Q. **How many brushes are there in a four pole (i) lap wound (ii) wave wound armature?**

Ans. (i) Four
(ii) Two.

Q. **What is pole pitch and coil pitch?**

Ans. The distance in slots between the centres of two adjacent poles is known as the pole pitch.

The coil pitch is defined as the distance in slots between the centres of two sides of a coil.

Q. **What is back pitch and front pitch?**

Ans. The back pitch is the distance between two sides of a coil at the back of armature. It is denoted by Y_B. The front pitch is the distance between two coil sides of a coil from front side. It is devoted by Y_F.

Q. **The distance between two adjacent coils which in other words is the resultant of back pitch and front pitch and shown as Y_R is known as what?**

Ans. Resultant pitch.

Q. **According to the movement of the winding, if the armature winding moves in the direction of the coil that is known as progressive winding but if it moves in the opposite direction, what do you call it?**

Ans. Retrogressive winding.

Q. **Can you classify the lap winding?**

Ans. Yes, it is simplex, duplex, triplex, quadruplex lap winding and multiplex winding.

Q. **What do you know about the multiplex winding?**

Ans. It is that type of winding in which two or more of simplex independent windings are done on the same armature.

Q. **What are the advantages of multiplex winding?**

Ans. Reduces the sparking, large current is also available, weight of the winding is also reduced, thin conductor can be used for more line current.

Q. **What are the classifications of wave winding?**

Ans. Simplex and multiplex, the multiplex includes duplex, triplex and quadruplex winding.

Q. **How brush are placed generally?**

Ans.
1. Parallel to the main field.
2. At right angle to the main magnetic field.
3. In between the magnetic field and magnetic neutral axis.

Q. **What are the methods of finding out the brush position?**

Ans. Ring method and degree method.

Q. **Name the apparatus used for testing the armature?**

Ans. Growler.

Q. **What are the names of the growler used for testing the armature of D.C. machine and A.C. stator winding?**

Ans. For D.C. machine—Outside growler.
For A.C. machine—Inside growler.

Q. **What are the types of A.C. winding?**

Ans. Basket winding, concentric winding, flat loop winding, chain winding, skew winding.

Q. **What are the types of coil generally used in A.C. winding?**

Ans. Mush coil (V shape and U shape coil), diamond coil and skew type coil.

Q. **What is skew winding?**

Ans. It is a hand winding, a very long coil is prepared and then kept in slot and after giving twists it is placed in other slot.

Q. **What is the coil group?**

Ans. A winding has a number of coil groups. In general the groups are equal to the number of poles.

Q. **What do you mean by coil connections and group connections?**

Ans. It is the connection which joins two coils of a group in series. It is also done as end to start or start to end.

Group connections are that connections in which a coil group is connected with other group.

Q. **What are the types of winding according to the layers, distribution and balancing?**

Ans. These are single layer, double layer, fully-distributed and partially distributed, balanced and unbalanced winding.

Q. **How the degrees per slot are calculated?**

Ans. First the number of poles are noticed, then total electrical degrees are known. Then the degree per slot can be calculated as

$$= \frac{\text{Total degrees}}{\text{No of slots}}$$

Q. **What are the tests to be performed for the winding?**

Ans. Continuity test, earth fault test, short circuit test, opposite and reverse connections and then balance test.

OBJECTIVE TYPE MULTIPLE CHOICE QUESTIONS

1. The D.C. armature winding is of the following types ......
   (a) lap and wave
   (b) series and parallel
   (c) series and shunt
   (d) commutating and compensating.

2. The wave winding has ...... number of parallel paths.
   (a) two (b) six
   (c) equal to number of poles (d) three.
3. The coils are divided in the followings, according to their construction ......
   (a) V-shape, U-shape, diamond type and skew types
   (b) V-shape, U-shape, lap and wave type
   (c) U-shape, diamond type and number of conductors
   (d) series, parallel and compound type.
4. The coil side which is ...... is known as active side.
   (a) outside the slot (b) inside the slot
   (c) on the commutator (d) on the segments.
5. The lap winding is of these following types ......
   (a) simplex, duplex, multiplex
   (b) simplex, triplex and multilayers
   (c) duplex, multistrand and multicolour
   (d) simplex, duplex, triplex and quadruplex.
6. The electric degrees are equal to 360° multiplied by the ......
   (a) number of parallel paths (b) number of brushes
   (c) number of pair of poles (d) number of poles
7. Generally for one h.p. three phase induction motor the ...... winding is used.
   (a) loop winding (b) concentric winding
   (c) basket winding (d) lap winding
8. Mostly the ceiling fans are wound with ...... type coils.
   (a) flat loop (b) concentric (c) diamond (d) skew.
9. In a 3ϕ, 24 slot, 12 coil 4 pole induction motor, the group connections are made ......
   (a) bottom to top (b) top to top
   (c) one to third coil (d) coils are not connected.
10. In star connection the similar ends are ......
    (a) connected (b) not connected
    (c) connected to earth
    (d) short circuited with starting terminals.
11. The distance between the two coil sides of a coil in terms of slots is known as ......
    (a) distribution factor (b) coil span factor
    (c) form factor (d) peak factor

12. The distance between top and bottom sides of a coil, measured around the back of the armature, is known as ......
    (a) back pitch (b) front pitch
    (c) commutator pitch (d) pitch factor
13. The distance between the starts of two consecutive coils measured in terms of slots is known as ......
    (a) back pitch (b) front pitch
    (c) winding pitch (d) commutator pitch
14. The coil which does not serve any electrical purpose is called ......
    (a) dummy coil (b) short circuited coil
    (c) unbalanced coil (d) distributed coil
15. The unequal values of emfs generated in different parallel paths give rise to resultant emf which acts across the armature winding. This unbalancing produces circulating current and losses even without load. To overcome this difficulty the ...... are used.
    (a) equalizer connections (b) group connections
    (c) sliprings (d) series connections
16. The winding in which the coil slots do not occupy all the circumference of the armature is known as ...... distributed winding.
    (a) partially (b) fully
    (c) half (d) None of the above.
17. When a reconnection method is adopted for a voltage change it must be kept in mind that voltage per coil or per turn be approximately if not exactly ......
    (a) same (b) half (c) double (d) 110 V.
18. The winding in which spiral coils are used is known as spiral winding. This winding is extensively used for armature winding of ......
    (a) alternators (b) A.C. motors
    (c) D.C. generators (d) rectifiers.
19. The chorded windings are preferred because it ...... the shape of sinusoidal waveform.
    (a) improves (b) deteriorates
    (c) cuts off (d) None of the above.
20. Which type of chorded winding is preferred?
    (a) Short pitched (b) Long pitched
21. Copper and labour cost is decreased in ......
    (a) short pitched winding (b) long pitched winding.
22. The whole coil winding has one coil group per pole per phase.
    (a) True (b) False.

23. The third harmonics of the waveforms have its frequency equal to ...... times the frequency of fundamental harmonic.
    (a) three (b) two (c) six (d) nine.
24. In the overhang the eddy current losses are negligible, therefore, the eddy current loss factor for the overhang is ......
    (a) 0.8 (b) 0.5 (c) unity (d) 0.01
25. If the number of coils per group are in whole number the grouping is said to be an even grouping, but if the coils per grouping are in fraction the grouping is called ......
    (a) phase grouping (b) even grouping
    (c) odd grouping (d) None of the above.

Answers To Objective Type Multiple Choice Questions

Chapter 1 Page 6

1. c	2. a	3. a	4. a	5. c
6. d	7. d	8. b	9. d	10. d

Chapter 2 Page 10

1. c	2. b	3. c	4. d	5. b
6. a	7. b	8. c	9. b	10. d

Chapter 3 Page 15

1. d	2. d	3. b	4. c	5. c
6. a	7. b	8. a	9. a	10. a

11. ohms, amp, watts, coulombs, kWh.
12. ammeter, voltmeter, frequency meter, p.f. meter, wattmeter
13. T, T, T, T, F.
14. energymeter, thermometer, megger, earth tester, voltmeter.
15. c

Chapter 4 Page 20

1. c	2. b	3. b	4. b	5. d
6. c	7. c	8. b	9. b	10. a
11. b	12. d			

13. phase wire, installation, hanging, D.B., neutral link.

Chapter 5 Page 29

1. b	2. a	3. a	4. a	5. a
6. d	7. b	8. b	9. a	10. b
11. b	12. c	13. a	14. d	15. a
16. c	17. a	18. c	19. b	20. b
21. c	22. b	23. b		

24. (a) increases (b) inversely
(c) opposite (d) sum
(e) increases (f) conductivity
(g) current × resistance (h) kWh or unit
(i) only one (j) ohm.

25. a	26. a	27. a	28. b	29. d
30. d	31. c	32. b	33. a	34. a
35. b	36. b	37. a	38. d	39. a
40. b	41. c	42. c	43. d	

Chapter 6 Page 36

1. a	2. a	3. d	4. a	5. a
6. c	7. c	8. c	9. b	10. b
11. c	12. b	13. a	14. b	15. b
16. a	17. a	18. c	19. c	20. d
21. d	22. c	23. a	24. b	25. c

Chapter 7 Page 49

1. a	2. b	3. b	4. d	5. a
6. b	7. c	8. b	9. c	10. a
11. b	12. a			

Chapter 8 Page 63

1. b	2. b	3. c	4. c	5. a
6. b	7. c	8. d	9. c	10. a
11. b	12. c	13. b	14. a	15. a
16. c	17. a	18. a	19. a	20. b
21. a	22. b	23. a	24. d	25. a

Chapter 9 Page 74

1. a	2. d	3. a	4. c	5. c
6. b	7. b	8. a	9. b	10. b
11. d	12. c	13. b	14. b	15. c
16. a	17. b	18. a	19. a	20. a
21. d	22. c	23. c	24. a	25. c
26. d	27. a	28. b	29. b	30. b

Chapter 10 Page 93

1. a	2. b	3. b	4. a	5. c
6. a	7. a	8. a	9. b	10. b
11. c	12. b	13. c	14. a	15. b

16. d	17. a	18. b	19. a	20. a
21. a	22. d	23. b	24. d	25. d

Chapter 11 Page 109

1. c	2. b	3. a	4. a	5. c
6. b	7. d	8. d	9. c	10. b

11. (a) F (b) T (c) T (d) F (e) F

12. (a) 2.75 V (b) 1.8 V
(c) 1280 (d) mud space
(e) separator (f) buckling
(g) boosting (h) gases
(i) D.C. (j) increases

13. (a) increases (b) chemical, electrical
(c) local action (d) amalgamated
(e) Daniell cell

14. (a) T (b) F (c) T (d) T (e) F (f) T (g) T (h) F (i) T (j) T

15. (a) negative with negative (b) No
(c) No in reverse direction (d) correct
(e) yes, varies from 10 to 20 mm.

16. a. 17 b. 21 (i) F (ii) T (iii) F (iv) T (v) T (vi) F (vii) T (viii) T (ix) F (x) F

Chapter 12 Page 122

1. c	2. d	3. a	4. a	5. a
6. c	7. d	8. c	9. c	10. a
11. b	12. a	13. b	14. b	15. d
16. d	17. a	18. c	19. d	20. a
21. c	22. d	23. d	24. a	25. d
26. d	27. a	28. a	29. a	

30. (a) F (b) T (c) F (d) T (e) T

Chapter 13 Page 139

1. c	2. b	3. b	4. a	5. a
6. c	7. d	8. d	9. a	10. a
11. a	12. d	13. b	14. c	15. b
16. b	17. b	18. a	19. c	20. d
21. c	22. c	23. a	24. b	25. a
26. b	27. a	28. a	29. a	30. d
31. a	32. a	33. b	34. a	35. b
36. b	37. a	38. b	39. c	40. a

41. b 42. d 43. b 44. d 45. c
46. d 47. d 48. a 49. c 50. c
51. c 52. a 53. a 54. b 55. d
56. a 57. d 58. c 59. d 60. d
61. a 62. b 63. b 64. a

65. (a) F (b) T (c) F (d) F (e) T (f) T (g) F (h) F (i) F (j) F

66. (a) T (b) F (c) F (d) F (e) F (f) T (g) T (h) T (i) F (j) T

67. (a) Rate of change of flux × No. of conductors
(b) Compensated winding (c) Rocker
(d) brushes (e) collect
(f) insulated (g) mica
(h) helps (i) soldered
(j) wave

68. a 69. a 70. d 71. d 72. a
73. c 74. (a) F (b) T (c) F (d) T (e) T

Chapter 14 Page 160

1. d 2. b 3. b 4. c 5. a
6. a 7. b 8. c 9. a 10. c
11. c 12. c 13. c 14. d 15. b

16. (a) T (b) T (c) F (d) F (e) F (f) F (g) T (h) F (i) T (j) T

17. b 18. a 19. c 20. b

Chapter 15 Page 176

1. a 2. a 3. a 4. b 5. b
6. b 7. b 8. d 9. c 10. c
11. c 12. b 13. c 14. c 15. d
16. a 17. a 18. c 19. a 20. c
21. b 22. a 23. a 24. a 25. c
26. a 27. c 28. a 29. a 30. a
31. a 32. c 33. a 34. c 35. a
36. b 37. d 38. d 39. d 40. c

41. (i) in phase, out of phase (ii) 90
(iii) vectorially (iv) resistance
(v) resonant circuit

42. (i) c/s (ii) no, ratio (iii) X_L (iv) watts (v) nothing (vi) minimum (vii) iron core (viii) resonant circuit (ix) ammeter and voltmeter

Chapter 16 Page 193

1. b 2. b 3. c 4. a 5. a
6. b 7. a 8. b 9. b 10. b

11. b	12. b	13. a	14. c	15. b
16. c	17. c	18. d	19. b	20. a
21. c	22. d	23. d	24. b	25. a

Chapter 17 Page 213

1. a	2. b	3. b	4. c	5. d
6. d	7. a	8. a	9. a	10. d
11. d	12. d	13. a	14. a	15. b
16. b	17. c	18. b	19. b	20. d
21. a	22. c	23. b	24. c	25. a
26. F	27. F	28. T	29. T	30. F

Chapter 18 Page 236

1. b	2. a	3. b	4. d	5. b
6. a	7. b	8. b	9. a	10. d
11. b	12. a	13. c	14. a	15. d
16. a	17. d	18. d	19. b	20. b
21. c	22. c	23. b	24. d	25. c

Chapter 19 Page 263

1. d	2. b	3. a	4. c	5. d
6. c	7. a	8. d	9. b	10. c
11. b	12. c	13. b	14. c	15. b
16. b	17. c	18. d	19. c	20. b

21. (a) Insulation resistance (b) uneven
(c) MΩ (d) 160 r.p.m.
(e) zero (f) infinity
(g) hand driven (h) 500 V
(i) 1000 V (j) low

22. (a) A.C./D.C. (b) uneven
(c) 10% (d) both
(e) two (f) air
(g) spring (h) ammeter and voltmeter
(i) ammeter (j) 270°
(k) insulated (l) ends

23. (a) for all as given (b) eddy current
(c) two (d) rotating
(e) uniform

24. c	25. c	26. c	27. d	28. a
29. a	30. a	31. b	32. c	33. a
34. b	35. c			

Chapter 20 Page 277

1. a	2. a	3. b	4. c	5. c
6. a	7. b	8. c	9. b	10. d
11. d	12. c	13. a	14. a	15. b
15. c	17. b	18. b	19. b	20. b

Chapter 21 Page 289

1. b	2. b	3. b	4. b	5. c
6. a	7. b	8. d	9. c	10. c
11. c	12. b	13. d	14. a	15. a
16. c	17. b	18. a	19. a	20. a

Chapter 22 Page 304

1. d	2. a	3. a	4. a	5. b
6. d	7. c	8. a	9. d	10. a
11. c	12. c	13. b	14. a	15. b

16. (i) T (ii) T (iii) T (iv) F (v) T (vi) T (vii) T (viii) T (ix) F (x) T

17. b	18. c	19. a	20. b	21. a
22. c	23. a	24. a	25. b	26. b
27. a	28. a	29. d	30. a	31. c
32. b	33. a	34. c	35. a	36. a
37. d	38. a			

Chapter 23 Page 311

1. c	2. c	3. a	4. c	5. c
6. b	7. b	8. d	9. b	10. c
11. a	12. a	13. a	14. a	15. a
16. d	17. d	18. c	19. c	20. c
21. b	22. a	23. b	24. c	25. a
26. d	27. b	28. d	29. a	30. d
31. b	32. b	33. a	34. a	35. a

Chapter 24 Page 317

1. a	2. a	3. a	4. b	5. d
6. c	7. c	8. a	9. a	10. a
11. b	12. a	13. c	14. a	15. a
16. a	17. a	18. a	19. a	20. a
21. a	22. a	23. a	24. c	25. c